THE GEOLOGY AND LANDSCAPE
OF SANTA BARBARA COUNTY, CALIFORNIA
AND ITS OFFSHORE ISLANDS

By

Robert M. Norris

Professor of Geology, Emeritus
University of California, Santa Barbara

SANTA BARBARA MUSEUM OF NATURAL HISTORY
2003

THE GEOLOGY AND LANDSCAPE OF SANTA BARBARA COUNTY, CALIFORNIA AND ITS OFFSHORE ISLANDS

Robert M. Norris

Santa Barbara Museum of Natural History Monographs Number 3

© 2003 Santa Barbara Museum of Natural History
2559 Puesta del Sol Road
Santa Barbara, California 93105-2936 USA
http:/www.sbnature.org

Library of Congress Cataloging-in-Publication Data

Norris, Robert M. (Robert Matheson), 1921-
The geology and landscape of Santa Barbara County, California, and its offshore islands / by Robert M. Norris.
p. cm. — (Santa Barbara Museum of Natural History monographs ; no. 3)
Includes bibliographical references and index.
ISBN 0-936494-35-2
1. Geology—California—Santa Barbara County. 2. Islands—California—Santa Barbara County. 3. Geology, Stratigraphic. I. Title. II. Series. QE90.S17 N67 2003 557.94'91—dc22
                    2003020476

© All rights reserved

This book may not be reproduced in whole or in part for any purpose whatsoever, without written permission from the publisher (Santa BArbara Museum of Natural History).

TABLE OF CONTENTS

PREFACE vii

ACKNOWLEDGMENTS ix

LIST OF ILLUSTRATIONS xi

INTRODUCTION xv

CHAPTER 1
Santa Barbara County's Place
    In the World 1
    In North America 2
    In California 4

CHAPTER 2
The Landscape of Santa Barbara County 7
Mountains 9
    Santa Ynez Mountains 9
    San Rafael Mountains 13
    Sierra Madre 16

Hills 17
    Santa Barbara Mesa, Hope Ranch, More Mesa and the Isla Vista area
    17
    Lompoc, Santa Rosa and White Hills 18
    Santa Rita Hills 19
    Purisima Hills 19
    Casmalia Hills and the Point Sal Ridge 20
    Solomon Hills including Graciosa Ridge, Camelback Hill and Gato Ridge 20

Streams and Valleys 21
    The Central Valley of Santa Cruz Island Cañada del Medio) 21
    Santa Ynez River Valley 22
    Los Alamos Valley 24
    Sisquoc River Valley 24
    Santa Maria and Cuyama Valleys 24
    Other Stream Valleys & Canyons 26

Lakes 30

Springs 32

Coastal Plains, Beaches and Elevated Marine Terraces 33
Coastal Erosion at Santa Barbara & Goleta 39

Sand Dunes 41

Headlands and Points 47
    Rincon Point 47
    Sand Point 48
    Loon Point 49
    Fernald Point 50
    Point Castillo 50
    Santa Barbara Point 51
    More Mesa Tar Deposit 51
    Goleta (Campus) Point 51
    Coal Oil Point 53
    El Capitan Point 55
    The Headland at Refugio Beach 55
    Point Conception 56
    Point Arguello 58
    Purisima Point 58
    Point Sal 59
    Mussel Rock 60

CHAPTER 3
The Rocks and Geologic History of Santa Barbara County 61

Jurassic and Older Rocks 69
    Santa Cruz Island Schist 69
    Franciscan Formation 69

Early Cretaceous Rocks 74
    Honda and Espada Formations 74

Late Cretaceous Rocks 74

Paleocene and Eocene Rocks 77
    Sierra Blanca Formation 78
    Anita Formation 78
    Juncal Formation 80
    Matilija, Cozy Dell and Coldwater formations 80
Oligocene Rocks 82

Sespe, Gaviota and Alegria Formations 82

Miocene Rocks 87
    The Vaqueros Formation 87
    Rincon Formation 89
    Monterey and Sisquoc Formations 90

Pliocene and Pleistocene Rocks 93
    Careaga Formation 94
    Santa Barbara Formation 95
    Casitas Formation 95
    Paso Robles 95
    The Fanglomerate 96

CHAPTER 4
Structural Features 97

Faults 97
    Santa Cruz Island and Santa Rosa Island Faults 99
    Santa Barbara Channel Structural Features 99
    Carpinteria Fault 100
    Arroyo Parida-Mission Ridge-More Ranch fault zone 100
    Mesa Fault 101
    Santa Ynez Fault 102
    Santa Ynez River Fault 103
    Little Pine Fault 103
    Camuesa Fault 104
    Big Pine Fault 104
    Rinconada Fault 106
    Ozena and South Cuyama Faults 106

Folds 107

Earthquakes 109

CHAPTER 5
Geologic Sketches of the Offshore Islands 113
    Santa Barbara Island 113
    Santa Cruz Island 114
    Santa Rosa Island 118
    San Miguel Island 120

CHAPTER 6
Mineral Resources 123
    Diatomite 123
    Metal Mining 126

Oil and Natural Gas 128

CHAPTER 7
Features of Special Interest 141
    Tecolote Tunnel 141
    Nojoqui Falls 142
    The Montecito Overturn 142
    Gaviota Gorge 142
    Figueroa Mountain's Black Smoker 143
    Guadalupe Dunes 144
    Growth of Mission Ridge, Santa Barbara 144
    Geologic Control of Vegetation 145
    The Sea Cliff at Isla Vista 146

ROAD LOGS
    U.S. Highway 101 from Rincon Creek to the Santa Maria River 149
    Refugio Pass Road from U.S. Highway 101 to State Highway 246 at Solvang 158
    West Camino Cielo from San Marcos Pass (State Highway 154) to Refugio Pass 160
    State Highway 1 from U.S. Highway 101 at Las Cruces to Guadelupe 162
    San Marcos Pass Route (State Highway 154) from US 101 in Santa Barbara to US 101 between Buellton and Los Alamos 166
    Lompoc to Honda Valley 169
    East Camino Cielo from Flores Flat Junction to Mono Forest Service Campground 170
    Gibraltar Road from Sheffield Reservoir to the Junction of East Camino Cielo Road with the San Marcos Pass Road (State Highway 154) ... 172
    Jalama Road from State Highway 1 to Jalama Beach County Park ... 176
    The Santa Ynez, Santa Rita and Lompoc Valley Route. State Highway 246 from State Highway 154 to Surf 178
    Paradise Road from State Highway 154 to locked gate near the SantaYnez River just beyond the Live Oak Forest Service Picnic Area 180
    Side Trip to Upper Oso Forest Service Campground 181
    Figueroa Mountain Loop 182
    Santa Maria, Sisquoc, Foxen Canyon, Los Olivos, Ballard Canyon, Solvang, and Nojoqui Falls 186

Table of Contents

Cat Canyon Road from Sisquoc via Palmer Road to U.S. Highway 101 188

Drum Canyon Road from Los Alamos to State Highway 246 189

State Highway 135 (San Antonio Road) From Los Alamos to State Route 1 via Vandenberg Air Force Base and Casmalia 190

Clark Avenue and Dominion and Palmer Road Loop 192

Harris Grade Route: Clark Avenue and U.S. Highway 101 via Orcutt to Lompoc 192

Across the Sierra Madre via Colson and La Brea Canyons and Miranda Pine Mountain 194

Santa Rosa Road from U.S. Highway 101 at Buellton to State Route 1 near Lompoc 197

Sweeney Road from State Highway 246 to End of the Road 198

Tepusquet Canyon from the Cuyama River Road to the Sisquoc River 199

Corralitos Canyon (Brown Road) State Route 1 to Point Sal Beach 201

The Cuyama River Valley Route from Santa Maria to the Junction of State Highways 166 and 33 202

Glossary 209

References 225

Index 229

PREFACE

When I was a graduate student in geology at UCLA following World War II, I and two fellow students mapped adjacent areas in the western part of Santa Barbara County, mostly on the San Julian, Hollister and Jalama ranches, as part of the requirements for the master degree. One of the ancillary benefits of this work was getting acquainted with the Dibblee family of the San Julian Ranch and the Hollisters of the Hollister Ranch on the south coast. Both families included well-known or to-be-well-known geologists, Joseph S. Hollister and Thomas W. Dibblee, Jr. Both helped me to understand the local rock sequence.

By 1948, when I was finishing up my work at UCLA, I innocently supposed that I had a reasonable familiarity with the geology of southern Santa Barbara County, but I never dreamed that I'd be spending my career teaching geology in the county. However, just four years later, I found myself the junior member of a two-man faculty in geology at what was then known as the Santa Barbara College campus of the University of California. One of my first duties involved organizing local geology field trips for the students. I soon found that my knowledge of the local geology was pretty thin, and though oil companies and Tom Dibblee had done a lot of work in the area, there was then not a great deal of published literature available. Local geologists, the scanty literature, and the efforts of my students whom I had put to work mapping in the area, greatly reduced my ignorance.

Despite the sometimes nearly impenetrable chaparral and abundant poison oak, the geology of the county proved to be varied and quite interesting, and the climate encouraged field work pretty much year around. My fellow faculty members in the field sciences such as biology and anthropology often had questions about local geology when it seemed to have some relevance to their work. Likewise, docents and field trip leaders at the Santa Barbara Museum of Natural History and the Santa Barbara Botanic Garden often came to me for some geologic information. From time to time they'd tell me that "someone" ought to prepare a layman's guide to the geology of the Santa Barbara area that would enhance their efforts. For quite a while I chose to think that the "someone" they had in mind wasn't me, but nobody rose to the challenge and eventually these suggestions from the docents became a bit more explicit, making it increasingly clear that the "someone" was me. In addition, I gradually became intrigued with the idea and realized it might become an enjoyable and possibly useful retirement project. Whatever the reason, it has encouraged me to review the now extensive published literature and maps, and to travel most of the county's roads and by-ways looking for features of geologic interest. I've enjoyed every minute of it, and I hope you, the readers, will find this little book useful and maybe even interesting. This book is not directed at the professional geologist, though any of that ilk that happen to read it are expected to let me know where and how I have erred.

ACKNOWLEDGMENTS

While writing this book, I often wished I knew the geology of the county in the intimate detail that Thomas W. Dibblee, Jr., knows it. Fortunately for me, and for you readers, Tom agreed to read one of the incarnations of the manuscript. In addition, the late Helmut Ehrenspeck, who has led numerous field trips all over the county, many with Tom Dibblee, also gave me the benefit of his considerable experience. I am grateful to these two geologists and am happy to concede that remaining errors that somehow escaped their attention are my responsibility. Alison Reitz salvaged a somewhat disorganized manuscript and rendered it more logical and easier to read. My son, Don, a professional editor, tidied up a number of awkward sentences and caught many typos previously missed. Special thanks to Eric Hochberg and Marie Murphy of the Santa Barbara Museum of Natural History whose editorial and design skills readied the manuscript for the printer. Wendy Bartlett, with a skillful and artistic eye, prepared the various line drawings and maps. Finally, my wife, Ginny, was an enormous help as well because she endured many bumpy, dusty roads, and late or missed meals as we assembled material for the road logs at the end of the book. Each of these people has my sincere appreciation and thanks.

Robert M. Norris
Santa Barbara, California
October 2003

LIST OF ILLUSTRATIONS

Figure 1	County place name location map	xvi
Figure 2	Geologic map of Santa Barbara County	3
Figure 3	Geomorphic provinces of Southern California	4
Figure 4	The Big Bend of the San Andreas fault	5
Figure 5	Geologic cross-sections, Santa Ynez Mountains	10
Figure 6	Geologic cross-sections, northern Santa Barbara County	11
Figure 7	Geologic cross-sections, offshore islands	12
Figure 8	Vertical, ripple-marked sandstone, Gibraltar Road	14
Figure 9	Drainage systems at Arroyo Burro and Laguna Blanca	16
Figure 10	Central Valley of Santa Cruz Island	22
Figure 11	Cuyama Valley badlands	26
Figure 12	Gaviota Pass	27
Figure 13	Aerial view of the Arroyo Burro area	29
Figure 14	Growth of the Mission Ridge	30
Figure 15	Zaca Lake	31
Figure 16	Operation of the longshore current	35
Figure 17	Beach sand transport cell	36
Figure 18	Aerial view of Santa Barbara Harbor in 1928	38
Figure 19	Aerial view of Santa Barbara Harbor in 1929	39
Figure 20	Aerial view of Santa Barbara Harbor in 1934	40
Figure 21	Aerial view of Santa Barbara Harbor in 1958	41
Figure 22	Dipping Monterey Formation at Gaviota Beach	42
Figure 23	Wave-cut shore platform at Santa Barbara Point	43
Figure 24	Guadalupe Dunes	44
Figure 25	Relative ages of sand dunes, western Santa Barbara County	45
Figure 26	Sand streams on San Miguel Island	46
Figure 27	Ghost Trees on San Miguel Island	47
Figure 28	Aerial view of the Rincon Creek area	48
Figure 29	Sand Point and El Estero, Carpinteria	49
Figure 30	Casitas Formation at Loon Point	50
Figure 31	Tar seep at More Mesa Beach	52
Figure 32	Drainage at Goleta Slough	54
Figure 33	Aerial view of the Goleta Point area	55

Figure 34	Boulder Delta at El Capitan State Beach	56
Figure 35	Point Conception from the sea	57
Figure 36	Point Sal	58
Figure 37	Pillow lavas at Point Sal	59
Figure 38	Geologic columnar section, Santa Cruz Island	63
Figure 39	Geologic columnar section, Santa Rosa Island	64
Figure 40	Geologic columnar section, Santa Ynez Mountains	65
Figure 41	Geologic columnar section, northern Santa Barbara County	66
Figure 42	Bedded Franciscan chert	70
Figure 43	Black Smoker at Figueroa Mountain	71
Figure 44	Diagram of Salinian Crustal black	73
Figure 45	Espada Formation, San Rafael Mountains	75
Figure 46	Crest of the Sierra Madre	77
Figure 47	Sierra Blanca limestone, Santa Ynez Mountains	79
Figure 48	Coldwater, Cozy Dell and Matilija formations, Santa Ynez Mountains	81
Figure 49	Lower Sespe Formation, Gibraltar Road, Santa Barbara	83
Figure 50	*Lyropecten magnolia* from the Vaqueros Formation	87
Figure 51	Vegetation break along the Rincon-Vaqueros contact, Santa Ynez Mountains	88
Figure 52	Celite diatomite quarry near Lompoc	91
Figure 53	Aerial view of elevated marine terraces near Santa Barbara	94
Figure 54	Pull-apart structure on the Carpinteria fault	101
Figure 55	Little Pine fault and Figueroa Mountain	105
Figure 56	Earthquake damage, Santa Barbara, 1925	110
Figure 57	Santa Barbara Island	114
Figure 58	Potato Harbor Formation, Santa Cruz Island	115
Figure 59	Blanca Formation, Santa Cruz Island	117
Figure 60	South coast of Santa Rosa Island	118
Figure 61	Microphotographs of diatoms from Sisquoc Formation	124
Figure 62	Active tar seep at Carpinteria State Beach	129
Figure 63	Landslide at Chinese Harbor, Santa Cruz Island	130
Figure 64	Summerland Oil Field about 1915	132
Figure 65	Hartnell No. 1 gusher, Orcutt Oil Field, 1904	133
Figure 66	Elwood Oil Field and the coastal marine terrace	135
Figure 67	Blowout at Platform A, Santa Barbara Channel, January 1969	136

Figure 68	Sea cliff erosion at Isla Vista, 1987	147
Figure 69	Sea cliff erosion at Isla Vista, 1993	147
Figure 70	Sea cliff erosion at Isla Vista, 1997	148
Figure 71	Call box sign on major highways	150
Figure 72	Paddle sign on State Highways	151
Figure 73	Cuyama River and the Caliente Range	203

INTRODUCTION

Santa Barbara County is not particularly large by California standards. And yet while it ranks 21st in size of the state's counties, it covers an area very nearly as large as Delaware and Rhode Island combined. Dubbed "California's Wonderful's Corner" by local writer and historian Walker Tomkins, Santa Barbara County occupies the western extent of the state's widest section. Three of California's 12 geomorphic provinces are represented: the southern Coast Ranges, the Transverse Ranges, and the Offshore Province (this includes parts of the western Continental Shelf and the southern California Continental Borderland). Moreover, half of southern California's eight islands fall within the county's boundaries.

Given its size, setting, mountainous character, and varied geology, Santa Barbara county presents a unique insight into many of the forces and processes that have shaped California (Figure 1).

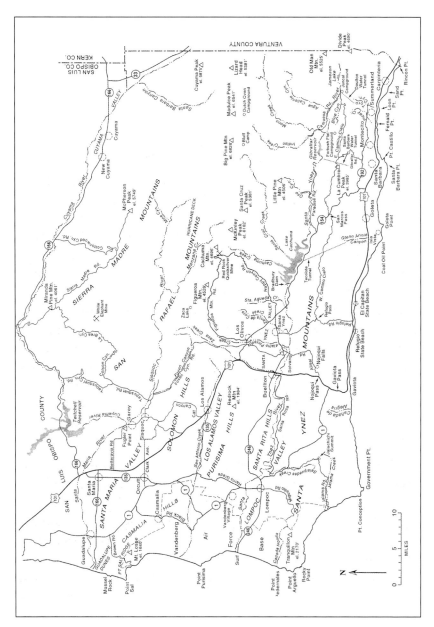

Figure 1 Map of Santa Barbara County showing place names and main roads.

CHAPTER 1

SANTA BARBARA COUNTY'S PLACE IN THE WORLD

In recent years we have learned that the entire outer crust of the earth is divided into a series of continent-sized blocks we call **plates**. It has also been found that the plates that make up the ocean floors form along great sutures like the Mid-Atlantic Ridge or the East Pacific Rise. We call these sutures **spreading centers** because they are sites where volcanic material wells up from the mantle below the crust. As it does, it adds material to the plates, forcing them to move away from the sutures. In the Atlantic Ocean, for example, this process is widening the Atlantic Ocean Basin and forcing the Americas away from Africa and Europe. But, these spreading centers also occur in other oceans, and since the earth is of finite size, widening cannot be going on in all ocean basins unless there is some counterveiling mechanism at work. As we have come to better understand **sea-floor spreading** and the associated process of **continental drift**, we have discovered some other sorts of plate boundaries that explain, for example, how spreading centers exist in both the Atlantic and Pacific basins.

The second type of plate boundary occurs where one plate is dragged or shoved beneath another. Because oceanic plates are thin and composed of relatively dense rock, where they impinge on continental plates that are thicker and made up of less dense rocks, the oceanic plate usually moves under the continental block, forming what we call a **subduction zone**. Considerable heat is generated during the subduction process and some rocks in the zone are melted while at the same time the overlying continental rocks are severely crumpled and elevated. As the sea floor near the continental edge is depressed over the descending oceanic plate, some of the molten rock in the subduction zone will rise through the edge of the overlying continental plate to form volcanoes in the coastal region. The west coast of South America is a splendid illustration of such a situation. The high Andean mountains have numerous active volcanoes and are located along the western edge of the continent. Just offshore, the descending oceanic plate has depressed the ocean floor producing the very deep Peru-Chile Trench.

As far as Santa Barbara is concerned, subduction died out here between 5-15 million years ago and was replaced by yet a third type of plate boundary, which, incidentally, explains why our local volcanic rocks are all about 10-15 millions years old.

About 15 million years ago, in our area, oceanic plate motion became more and more oblique with respect to the western margin of North America. By about 5 million years ago, subduction was replaced by the third type of plate boundary in which the two plates slide by one another. This type of boundary is called a **transform fault** and our Californian example is the great San Andreas fault that extends from the Salton Sea northwestward to Cape Mendocino. Total slip on this fault since it became a plate boundary now amounts to about 350 miles in which the block on the western side has moved northwesterly with respect to the continent on the east.

During the transition from subduction to transform faulting, the Pacific Plate detached a narrow, elongate strip of the North American Plate. This strip includes all of Santa Barbara County, all of the Baja California peninsula and all of coastal California as far north as San Francisco. The movement of the Pacific Plate is thus carrying our part of the continent northwesterly at about 1-2 inches a year. Should this motion continue at the present rate, it will bring the Los Angeles area adjacent to the San Francisco Bay area in about 11 million years.

IN NORTH AMERICA

Most geologists now agree that the motions of these large crustal plates account for the chain of mountains all along the western edge of the Americas from Alaska to Tierra del Fuego, including all the mountains of Santa Barbara County. The North American part of this crumpled, complex chain of mountains is collectively called "The Western Cordillera". There is abundant evidence not only in the form of earthquakes and bent, broken and tilted rocks of young geologic age that this Cordilleran belt is still being actively deformed. Hence, mountain building activity continues in Santa Barbara County.

Geologically recent events show that we live in one of the more rapidly changing environments in North America and because of this, geologists describe the county's landscape as very young. This is reasonable because landscape is the product of two opposing activities or processes. Plate motions elevate, deform, or break (fault) rocks and generate volcanic activity, giving rise to mountains and elevated plateaus.

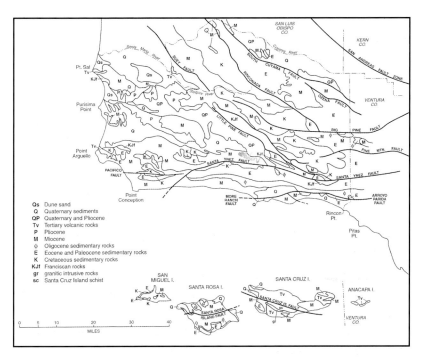

Figure 2   Geologic map of Santa Barbara County.

Erosional processes, on the other hand, destroy highlands mostly by running water, but sometimes also by glacial activity and even wind erosion. These processes transfer material from the highlands to the lowlands, fill basins and valleys, while carving canyons and gorges in the uplands. The steeper and higher the mountains, the faster the streams cut their canyons. Over time, if the pace of uplift slows, the landscape becomes "older" and mountains are reduced to hills, valleys fill and widen and the amount of relief diminishes. By relief we mean the difference in elevation between the highest and lowest places in a given landscape.

As far as our landscape in Santa Barbara County, it is still a very youthful one in which mountains dominate and erosion has only begun to flatten the landscape. Mountain canyons remain steep and narrow and features formed by faulting have not yet been eroded away or buried, but remain fresh and prominent. This is because few, if any, of our local landscape features are more than 2-3 million years old–only a fleeting moment in the vast stretch of geologic time. On the other hand, from a human perspective, two million years is a very long time indeed and the processes of uplift and erosion can seem lethargically slow. Even notable earthquakes such as occurred in 1925, 1941, 1952 and 1978

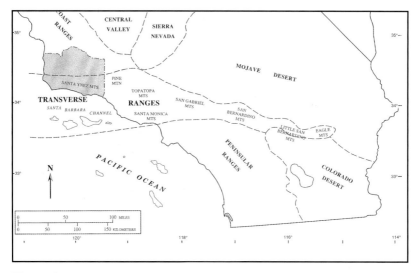

Figure 3   Geomorphic provinces of southern California.

may be seen as relatively infrequent, but at that rate, there would be 40,000 earthquakes every million years. If each quake was accompanied by only a few inches of displacement, in a million years there would be rather large changes.

IN CALIFORNIA

All of California lies within the Western Cordillera of North America, but within the state there are 12 smaller, but well-defined geomorphic provinces or regions. These are based chiefly on landforms and 11 of the 12 are on land; the twelfth is the offshore area made up of two parts, the narrow **Continental Shelf** north of Point Arguello and the very irregular area of deep basins and elevated ridges and islands called the **Southern California Continental Borderland** south and east of the three offshore islands, Santa Cruz, Santa Rosa and San Miguel.

Familiar examples of some of the geomorphic provinces on land include the backbone of California, the 400-mile long Sierra Nevada, the Great or Central Valley (a 450-mile long nearly flat depression lying between the Sierra Nevada and the Coast Ranges,) and the unusual east-west-trending Transverse Ranges of southern California. With few exceptions, such as the mountains of central Alaska, the Uinta Range of Utah and the low mountains

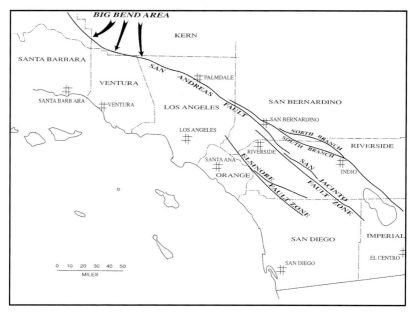

Figure 4.   Map of southern California showing the location of the San Andreas and other major faults and the Big Bend area.

of southern Oklahoma, all the mountains in North America have a general north-south trend.

In Santa Barbara County, the Transverse Ranges include the Santa Ynez Range, the Santa Barbara Channel and the three offshore islands, San Miguel, Santa Rosa and Santa Cruz. Anacapa Island belongs in this group as well, but is in Ventura County. The province extends east embracing the mountains of Ventura and Los Angeles counties as well as those of southern San Bernardino and northern Riverside counties.

Not surprisingly, the unusual trend of the Transverse Ranges is attributed to plate interaction. The San Andreas fault veers sharply west where it enters the Transverse Ranges (Figures 2 and 3), forming what is called the "Big Bend". The northwesterly movement of the block on the seaward side of the fault forces it against the continental block in the Big Bend area, shortening the crust there and producing an east-west belt of mountains and valleys (Figure 4).

CHAPTER 2

THE LANDSCAPE OF SANTA BARBARA COUNTY

Before discussing the mountains and valleys of the county, it should be noted that a geologist will often assert that the mountains are "young", but then seemingly confuse the matter by reporting that the rocks making up the youthful mountains are in fact hundreds of millions of years old. We must be careful to distinguish the age of the landscape from the age of the rocks that comprise it. It's a bit like saying that a modern brick wall is young, even though it is made of bricks from a Roman ruin. Similarly, someone might observe that rocks high in the mountains were remains of an ancient sea bed. Does the fact that they are now thousands of feet above sea level mean that the oceans once covered the mountains? The answer in this case is "No", and here's why:

A large share of the rocks exposed in Santa Barbara County are sedimentary and were originally loose sand, soft mud or gravel deposited on the sea floor, sometimes near shore but also in very deep basins. In addition, some of the sediments were deposited on land in streams or lakes. Fossils preserved in these sedimentary rocks can tell us much about the environment of deposition, whether the water was deep or shallow, warm or cold, and what the climate was like.

Most of the mainland part of the county has a landscape developed on sedimentary rocks. The other two large classes of rock, igneous and metamorphic, are largely confined to a belt of rock in the vicinity of Figueroa Mountain and at Point Sal. Volcanic (igneous) and metamorphic rocks are considerably more important in the landscape of the offshore islands.

The landscape developed on igneous and metamorphic rock is, as a rule, only subtly different from landscape forming on sedimentary rock. Erosional processes respond more to rock resistance than to rock class. For example, sandstone layers in the Santa Ynez Mountains are appreciably more resistant to erosion than the softer shales and mudstones with which they are associated. As a result, the prominent ridges of bare rock are sandstones whereas the adjacent layers of shale and mudstone form swales between the hard rocks and give rise to smoother, less craggy slopes.

The belt of metamorphic and igneous rocks that extends southeast from the Figueroa Mountain area has a distinctive hummocky topography because the rocks of the Franciscan Formation exposed there include masses of slippery serpentinite (which is subject to landsliding) along with resistant masses of hard chert and volcanic rock enclosed in softer, weaker sandstones called graywackes.

As a result of this mixture of rock types, the hillsides are often dotted with numerous irregular knobs and hillocks composed of the more resistant rock types. These are known as "knockers" to geologists, and are a prominent feature in the California Coast Ranges from Santa Barbara County north to the Oregon border.

Sometimes the erosional agent itself produces a distinctive landscape. The glacial landscapes in Yosemite Valley in the Sierra Nevada are good examples, but glaciers have played no role at all in forming the landscape of Santa Barbara County. However, wind as an erosional and depositional agent has produced some of the county's more distinctive landscapes. Perhaps surprisingly, there are no good examples in the driest part of the county, the upper Cuyama Valley, but there are many clear instances along the west-facing coast north of Point Arguello and on the offshore islands where large fields of wind-deposited sand dunes occur.

Finally, all of the world's shores are affected by wave erosion and deposition. Wave-modified landscapes are prominent in Santa Barbara County and its islands because our mountainous terrain encourages development of prominent seacliffs along much of our shoreline.

Waves attack shorelines somewhat like a horizontal saw operating between the tidal limits. If the land is elevated abruptly, perhaps by faulting, or if sea level declines rapidly, a well-preserved beach and sea cliff may be left stranded above the present shoreline. It is rather common along our mountainous coastline to find places where three or four elevated platforms and seacliffs can be seen. A few of these wave cut benches or terraces are a mile or more wide, though most are much narrower.

It is clear then that virtually all erosional features that make the county's landscape, apart from those along the immediate coast, are attributable solely to the effects of running water, whether it be the badland landscapes of the arid Cuyama Valley, the broad Santa Ynez River Valley or the narrow Gaviota Gorge. This is true despite the fact that many, perhaps most of the county's streams are intermittent and carry water only during and shortly after rains.

Chapter 2

MOUNTAINS

Most of the land area of Santa Barbara County, including the islands, is either hilly or mountainous, an indication that the landscape is quite young, geologically speaking. Although erosional processes have carved a myriad of canyons and gorges into these uplands, the processes that together have raised the mountains and hills continue to outpace the destructive effects of erosion. We will be giving a number of examples that show that faulting, folding and uplift remain active as we consider the various landscape features of the county.

All the mountains of the county involve both folding and faulting as illustrated in the geologic cross-sections (Figure 5 to 7), but some of the hilly areas in the western part of the county are chiefly folded features with only minor faulting.

Santa Ynez Mountains

This east-west trending range has a single, well-defined crest and extends almost 70 miles from western Ventura County near Ojai, to Point Conception. Its steep south front faces a narrow coastal plain and the Santa Barbara Channel. West of San Marcos Pass, the northern side of the range is bold and prominent as far as Gaviota Pass, but to the east, it increasingly merges with the southern Coast Ranges, losing its distinctive character by the time the Ventura County line is reached.

By far, the majority of rocks in the Santa Ynez Mountains are sedimentary and are Cretaceous or early Tertiary in age. There are some volcanic rocks in the westernmost part of the range in the vicinity of Tranquillon Mountain, and an area of Franciscan metamorphic rock between Gibraltar and Jameson lakes. Resistant rocks, mostly sandstones, form the prominent, often bare ridges along the crest and southern face of the mountains.

North-south crustal shortening or compression has raised and tilted this range along the Santa Ynez fault that defines the northern margin of the mountains, forming a ramp on which the range has been shoved upward and tilted. This crustal shortening is also responsible for creating the whole of the Transverse Range province that raised the Santa Ynez Mountains and the Channel Islands and depressed the intervening Santa Barbara Channel. During uplift of the Santa Ynez Mountains, its rock layers were tilted downward toward the sea forming what geologists call a **homocline**, that is, a stack of rocks all tilted or dipping in the same direction.

10 The Geology and Landscape of Santa Barbara County, California

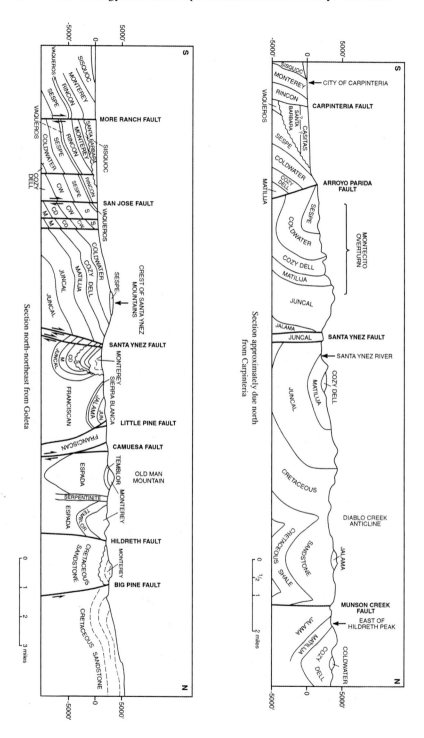

Figure 5. Geologic cross-sections of the Santa Ynez Mountains. (After T. W. Dibblee, 1986, 1987).

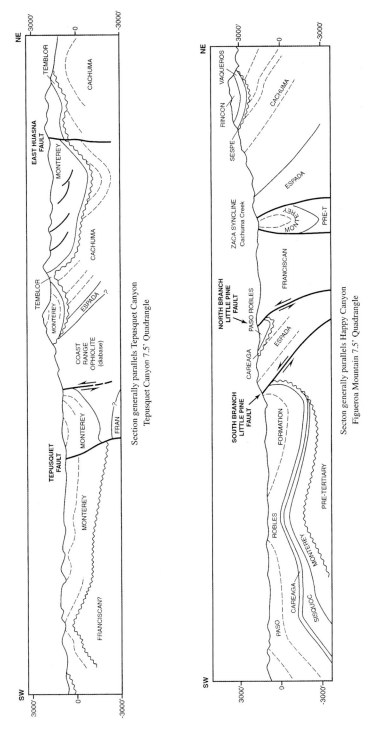

Figure 6. Geologic cross-sections of northern Santa Barbara County. (After T. W. Dibble, left-1994 & right-1993).

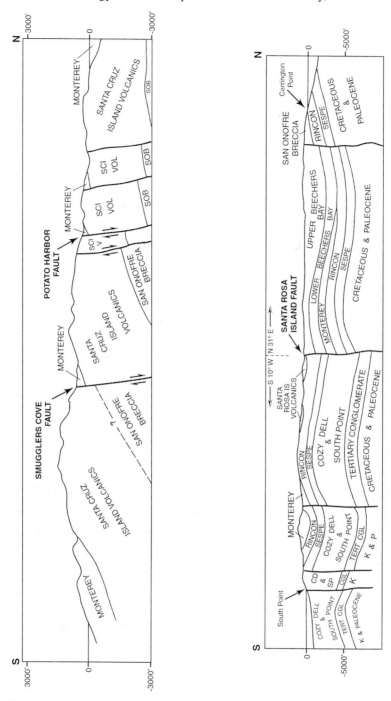

Figure 7. Geologic cross-sections of the Northern Channel Islands. Left: eastern end of Santa Cruz Island; right: central Santa Rosa Island. (Left after D. W. Weaver *et al.*, 1969 & AAPG Pac. Sect. Pub. MP 46, 1998; right after P. W. Weigand, 1998).

However, when one examines the Santa Ynez Mountains more closely, it becomes evident that within its generally homoclinal nature, there are a number of variations. Among these is a syncline easily seen from San Marcos Pass Road (see Road log for State Highway 154). This fold cuts diagonally across the range near San Marcos Pass and is responsible for the saddle crossed by the highway.

Eastward from San Marcos Pass, the inclination of the rock layers on the south face of the range becomes progressively steeper, a result of locally more intense compression or squeezing as the Transverse Ranges developed. On Gibraltar Road north of the city of Santa Barbara, some of the rocks are vertical (Figure 8), and if followed eastward toward the Ventura County line, become overturned and no longer dip seaward, but instead northward into the mountains. This feature is called the "Montecito overturn" and is illustrated on the cross-section for the Carpinteria area (Figure 5).

The crest of the Santa Ynez Mountains and the steepness of its south face reach a maximum near the county line in the Montecito Overturn area where the highest points just exceed 4600 feet. At San Marcos Pass, in contrast, the crestal elevation is only about 2200 feet as a result of the synclinal fold that lies athwart the axis of the mountains there, despite the fact that the rocks at the summit belong to the resistant Coldwater sandstone. Westward, the elevation of the crest again increases to about 4200 feet at Broadcast and Santa Ynez peaks. Both of these peaks are formed on the relatively durable Matilija sandstone.

Farther west, Refugio Pass, like San Marcos Pass, has an elevation of about 2200 feet, but unlike the latter it lies on the less resistant Sacate Formation. The Sacate Formation is much more easily eroded than either the massive Coldwater sandstone at San Marcos Pass, or the durable Matilija at Broadcast and Santa Ynez peaks.

West of Gaviota Pass, the range becomes appreciably lower, seldom exceeding 1500 feet except at its far western end near Tranquillon Mountain where the crest barely exceeds 2000 feet. The volcanic rocks at this peak are somewhat more resistant to erosion than the surrounding sedimentary rocks.

San Rafael Mountains

This is a multi-crested complex of mountains forming the highest and southernmost unit of the California Coast Ranges. It lies between two single-crested ranges, the Sierra Madre on the north and the Santa Ynez to the

Figure 8. Gibraltar Road near Santa Barbara. Nearly vertical sandstone bed in the Eocene Cozy Dell Formation showing shallow-water ripple marks.

south, but merges with both of these ranges toward the east, making any boundary selected in that area arbitrary and hard to define. The San Rafael Mountains occupy the transitional area where the northwest-trending Coast Ranges curve around and join the east-west-trending Transverse Ranges.

Big Pine Mountain (6828 ft) is the highest point in the county and also the highest point in the California Coast Ranges south of the Clear Lake area.

Few public roads extend very far into the San Rafael Mountains except in the Figueroa Mountain area and in the Colson and La Brea Canyon area east of Santa Maria. A considerable portion of this mountain complex has been allocated to the San Rafael and Dick Smith roadless wilderness areas.

The most distinctive rock unit in the San Rafael Mountains is the varied Franciscan Formation that forms a belt from the Figueroa Mountain area southeastward to near Gibraltar Reservoir. It is described in the chapter on rocks as well as in the Figueroa Mountain road log. Franciscan rocks are at least partly of Jurassic age and as such are the oldest rocks in the mainland part of the county. Marine sedimentary rocks of late Cretaceous and early Tertiary age make up the major part of exposed rocks in the San Rafael Mountains, but are overlain in many places by the ubiquitous Monterey Formation of Miocene age. Most of the high peaks are on Monterey rocks and perhaps surprisingly most are synclinal folds. One might expect that downfolds or synclines to be sites of valleys, and **anticlines** to be characteristic of the higher areas. One reason that anticlines seldom form highlands is because the rocks across the tops of these folds are stretched and fractured during folding whereas the rocks in the center of a syncline are squeezed and compressed. The broken crests of anticlines are therefore prone to fall prey to erosional processes and are cut away more rapidly than the adjacent synclinal folds. Local differences in rock resistance may, of course, play a role as well.

Examples of mountains or high ridges developed on synclinally folded rocks of the Monterey Formation are Figueroa Mountain, Zaca Peak, Wheat Peak, and Little Pine Mountain. The long nearly straight ridge that stretches from Wheat Peak and Bald Mountain southeastward, called the Hurricane Deck, is a synclinal fold in Monterey rock.

The name "Hurricane Deck" suggests a windy tableland or mesa, but it is in fact a fairly narrow, roughly straight ridge about 10 miles long. This ridge forms the southern rim of the upper Sisquoc River Valley and has an elevation of about 4000 feet. Its windiness, rather than its topography is the probably source of its name.

Most of the other high peaks in the San Rafael Mountains, such as Big Pine Mountain (6828 ft), Hildreth Peak (5065 ft), Santa Cruz Peak (5381 ft), McKinley Mountain (6182 ft), and Cuyama Peak (5875 ft), are all located on late Cretaceous to early Tertiary marine pebbly sandstones, generally rather monotonous and not very fossiliferous; the same sort of rocks also form the crest of the Sierra Madre.

16   The Geology and Landscape of Santa Barbara County, California

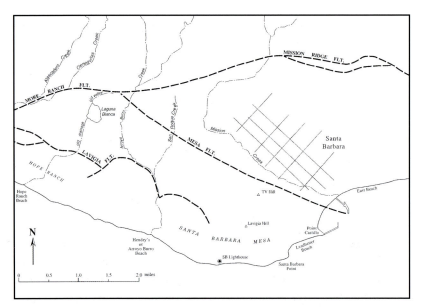

Figure 9.   Map of the area between Santa Barbara and Hope Ranch showing the location of drainage patterns and major faults.

The part of the San Rafael Mountain complex that lies northwest of Tepusquet Canyon is sometimes regarded as part of the Sierra Madre, particularly that part south and west of Los Coches Mountain (3016 ft). This entire area is almost an unbroken exposure of the Monterey Formation.

Sierra Madre

For our purposes, we limit the Sierra Madre to a reasonably well-defined single ridge stretching about 30 miles southeast from near Miranda Pine Mountain (4061'), to the area of Salisbury Potrero. Wherever the boundary is drawn, the Sierra Madre certainly merges on the southeast with the San Rafael Mountains.

The bulk of rocks in this range are a rather unremarkable series of yellowish-brown marine sandstones with minor amounts of pebbly conglomerate and shale. Fossils are not very common, but those that have been found give an age ranging from late Cretaceous to early Tertiary (Paleocene).

Along the northeastern base of the mountains, facing the Cuyama River Valley, are patches of white, marine Santa Margarita sandstone of late Miocene age. In some places the Santa Margarita is very fossiliferous, and contains an abundance of sand dollars. After the Santa Margarita was deposited in this area, the sea withdrew and all the younger rocks such as the Morales Formation were formed by streams or sometimes in lakes.

HILLS

Santa Barbara Mesa, Hope Ranch, More Mesa and the Isla Vista area

Westward from the city of Santa Barbara are the hills of the Mesa district, Hope Ranch, the flat-topped More Mesa and finally the nearly level modestly elevated marine terrace at UCSB and Isla Vista.

This belt of hills has been raised along the More Ranch fault on the north and the Mesa fault on the east (Figure 9). This uplift has been anything but uniform — Isla Vista and the UCSB campus are about 40 feet above sea level, More Mesa about 100, Hope Ranch as much as 600 and Santa Barbara Mesa about 400 at its highest point. Several benches or marine terraces have been cut into the hills of Hope Ranch and at Santa Barbara Mesa, but More Mesa and the Isla Vista area are mainly single uplifted benches of marine erosion. Good dates have been obtained on the formation of these marine terraces. The 40-foot terrace at Isla Vista is about 40,000 years old (late Pleistocene), but the 40,000 year old terrace at Hope Ranch is about 200 feet above sea level.

The bedrock in these hills and elevated marine terraces belongs to three marine formations. The oldest is the Monterey Formation of Miocene age and is best seen in the seacliffs from Santa Barbara Harbor westward to More Mesa Beach, as well as in the cliffs that form the eastern side of the Goleta Slough entrance. The slightly younger, but also Miocene Age, Sisquoc Formation is exposed in the seacliff at both ends of More Mesa Beach as well as in the cliffs along the UCSB campus and at Isla Vista.

As a rule, the younger Santa Barbara Formation makes up the northern side of these hills and is, in the central part, separated from the older Monterey Formation by the Lavigia fault (Figure 9). At More Mesa, however, a gentle downwarp has depressed both the Sisquoc and Monterey formations a bit below sea level, so that the high cliffs along that beach are formed from the Santa Barbara Formation that rests on these older rocks.

Both the deformation of the young marine terraces and the faulting of the slightly older Santa Barbara Formation testify to the geologically youthful age of these hills. Further deformation by faulting or folding would certainly be no surprise to a geologist, nor should it be to any well-informed resident of the Santa Barbara area.

Lompoc, Santa Rosa and White Hills

West of Gaviota Pass the Santa Ynez Range retains its single crest as far west as Point Conception, but it generally has an elevation less than 1500 feet. This contrasts sharply with the much higher range crest east of Gaviota, especially when viewed from its seaward side.

The northern side of the range west of Gaviota is not nearly so well-defined as it is east of that pass, but instead merges with hilly areas lying south of the Santa Ynez River. These hilly areas are, from east to west, the Santa Rosa, the Lompoc and the White hills and on the west, Tranquillon Mountain near Point Arguello. Although these local names are well established, there is a good reason to consider them all parts of the western Santa Ynez Mountains because together they form a continuous group of hills and low mountains.

Apart from Tranquillon Mountain, the rocks in these hills are chiefly marine sedimentary deposits ranging in age from late Cretaceous to Miocene. In common with the main crest of the Santa Ynez Mountains on the south, faults and folds have generally east-west trend. Folding is much more prominent than faulting in the hilly areas north of the system of faults that lies along the northern side of the Santa Ynez Range.

The southern parts of the Santa Rosa and Lompoc hills along State Highway 1, El Jaro Creek and in the Salsipuedes Creek drainage, include many exposures of late Cretaceous and early Tertiary marine sedimentary rocks, but the northern and western parts of these hills are underlain almost entirely by the Miocene Monterey and Sisquoc formations.

The White Hills south of Lompoc take their name from the white diatomite beds of the Sisquoc Formation exposed there. A large commercial diatomite quarrying operation is located in these hills.

At Tranquillon Mountain (2170 ft), inland from Point Arguello, is the most extensive exposure of volcanic rocks anywhere in the mainland part of the county. These are mostly rhyolites, but include some andesite and basalt, darker-colored and more iron-rich than the rhyolites. Most of these rocks are flow-breccias that formed as nearly-congealed, pasty, probably thick lava flows advanced slowly over the surface, breaking up into large, angular blocks in the process. Some of these flow breccias were deposited on land and others on the sea floor. In addition, we know that these were accompanied by explosive eruptions because ash beds are present among the flows, and also in the closely associated marine Miocene Monterey Formation. The Tranquillon flows and ash beds are the same age as the enclosing Monterey Formation. The Obispo Tuff in the north county near Twitchell Dam and the volcanic rocks on the Channel Islands are also about the same age as the Tranquillon volcanics. The Tranquillon volcanic rocks are just one example of the widespread volcanic activity that affected most of California during Miocene time.

Santa Rita Hills

This group of hills lies between the Santa Rita Valley on the north through which State Highway 246 passes, and the Santa Ynez River to the south. These hills result from a broad anticlinal or arch-like fold. Erosion of the crest of the fold has exposed the Monterey Formation in its core, but the outer flanks of the hills are made up of younger marine rocks belonging to the Sisquoc and Careaga formations. Here and there, along the lower flanks of the hills are patches of the still younger, non-marine Paso Robles Formation.

Purisima Hills

The Purisima Hills extend from U.S. Highway 101 on the east, to State Highway 1 on the west. They are bounded on the north by the Los Alamos Valley and on the south by the Santa Rita Valley and State Highway 246.

Like the nearby Santa Rita Hills, the Purisima Hills are anticlinal, but only in the vicinity of Redrock Mountain (1964 ft) at the eastern end of the hills, is the Monterey Formation exposed; nearly everywhere else in the hills, exposed rocks are either the Careaga or Sisquoc formations. Some of the better exposures of these two rock units are along the two roads that cross the axis of the hills, Drum Canyon on the east, and the Harris Grade road on the west.

The Lompoc Oil Field is located on the south flank of the Purisima Hills both east and west of the Harris Grade Road. This field was discovered in 1903 by the Union Oil Company. It remains a small producer of heavy oil. One of the wells in this field set something of a record by flowing without pumping from 1903 to 1922.

Casmalia Hills and the Point Sal Ridge

The Casmalia Hills could be considered the northwestern extension of the Purisima Hills because they are separated only by the valley of San Antonio Creek. Southeast of the town of Casmalia, the hills are in most respects quite similar to the Purisima Hills; they are broadly anticlinal and have a similar set of sedimentary rocks. Northwest toward Point Sal, however, the Sisquoc and Careaga formations are largely absent and the older Monterey Formation forms the crest of the hills almost to Mount Lospe (1640 ft).

A distinctly different set of rocks is present in the Point Sal Ridge west of Mount Lospe. Among these are the red, non-marine Lospe Formation of late Miocene age and the very distinctive deep-sea sequence of volcanic and sedimentary rocks called ophiolite. Best exposures of this unusual set of rocks are to be found along the shore east and north of Point Sal. These rocks and their origin are more fully described in the chapter on rocks, as well as in the Brown Road-Point Sal road log.

Northwest of Corralitos Canyon, the oldest of the widespread dune deposits (the Orcutt sand) laps up on the flanks of the Casmalia Hills. It is in this area that this blanket of old dune sand reaches its maximum elevation of just over 1100 feet, completely covering the older rocks beneath.

The small Casmalia Oil Field is located on an anticlinal fold between Orcutt and Casmalia. This is another old oil field, in production since 1917. It yields a thick, tarry oil used mainly for paving and sealing purposes.

Solomon Hills including Graciosa Ridge, Camelback Hill and Gato Ridge

This group of hills is bounded on the northeast by the Sisquoc River and Santa Maria valleys, and on the southwest by the Los Alamos or San Antonio Creek valleys. The highest point in the hills is the hat-shaped Mount Solomon (346 ft), now bristling with telecommunication towers and located just a short distance west of U.S. Highway 101.

Like the other hills to the south, the Solomon Hills are primarily the result of folding, though there are some minor faults present. The folding is somewhat more complex here than in the other hilly areas, and consists of three parallel anticlinal folds separated by two down-warped or synclinal folds. U.S. Highway 101, from near its junction with Cat Canyon Road, to Mount Solomon follows one of these synclines. The Solomon anticline lies just southwest of the highway here and the Cat Canyon or Flores anticline lies to the northeast. Still farther northeast and roughly parallel to the Solomon and Cat Canyon anticlines is the Gato Ridge anticline.

All three of these anticlinal folds produce oil. The Orcutt field lies on the Mount Solomon anticline, the Cat Canyon field on the Cat Canyon anticline and the East Cat Canyon and several associated smaller fields on the Gato Ridge anticline. All three of the main fields came into production between 1902 and 1909 and were still producing oil in November 2000.

The oldest rocks seen anywhere in the Solomon Hills are the marine sandstones of the Pliocene Careaga Formation. A large part of the hills is covered with the younger Paso Robles and Orcutt formations, both of Pleistocene age. (See road logs for Cat Canyon, Clark, Dominion and Palmer roads).

STREAMS AND VALLEYS

Santa Cruz Island's Central Valley (Cañada del Medio)

Starting from the south, perhaps the most unusual valley in the county is the Central Valley of Santa Cruz Island, a feature 12.5 miles long eroded by stream action along the Santa Cruz Island fault. A small portion of this valley at both its eastern and western ends drains directly into the sea, but most of the valley drains through a winding gorge into Prisoners Harbor on the north coast. This would seem an unlikely course because it would appear much more logical for the stream to follow the open valley to the east where it would have to cross only a very low divide to reach the sea instead of crossing the much more formidable mountain barrier between the Central Valley and Prisoners Harbor. It is probable, but not at all certain, that this anomalous stream established its course before the mountains on the north side of the valley were elevated to their present height, but to date no one has studied the question very carefully (Figure 10).

Santa Ynez River Valley

One of the two most prominent valleys in the county is the Santa Ynez River Valley that extends 75 miles from the Ventura County line near Old Man Mountain, westward to the coast at Surf, west of Lompoc. The upper part of this valley near Juncal Dam is steep and narrow, but it widens near Lake Cachuma to form a nearly flat, more or less triangular plain called the

Figure 10. Central Valley, Santa Cruz Island. View east down the valley that was eroded along the Santa Cruz Island fault system. (Photo by E. W. Keller).

Santa Ynez Valley with its corners at Los Olivos, Solvang, and where State Highway 154 crosses the Santa Ynez River (Figure 1). The floor of the valley is gently tilted to the south and is made up of a blanket of stream deposits, sands and gravels, laid down by southward-flowing Alamo Pintado Creek on the west, Zanja de Cota Creek in the center and Santa Agueda Creek on the east. The Santa Ynez River and its flood plain deposits form the southern side of this triangle.

Downstream from Solvang, the Santa Ynez River Valley narrows appreciably, but opens again near Lompoc, where the valley is almost four miles wide. The rich agricultural soils of the Lompoc area are Santa Ynez River flood plain deposits.

Upstream from Lake Cachuma, the river generally follows the Santa Ynez fault because fractured rock along the fault is more easily eroded than unbroken rock on either side of the valley. Downstream, the course of the river and the location of the valley have developed for other reasons. For example, the river leaves the broad Santa Ynez Valley through the narrow gorge (Lompoc Narrows), suggesting that, like the drainage on Santa Cruz Island, the Santa Ynez River had established its course before the present day hills at the Lompoc Narrows were raised in its path.

The three dams on the river, Juncal on the east, Gibraltar next and Bradbury on the west, have, of course, greatly altered its flow. Rainfall is highly variable from one year to the next in southern California and the typically long, dry summers result in widely fluctuating flows. The water agencies that control the three reservoirs on the river generally prefer to release water downstream only when the reservoirs are full or nearly full. The small upstream lakes, Jameson and Gibraltar, will fill in most winters, but the largest reservoir, Lake Cachuma, may fill only once in five or six years and seldom spills two years in a row. More frequent releases of water are often necessary to provide for the downstream needs of Solvang, Buellton and Lompoc.

Some reaches of the river commonly go dry at the end of the summer and almost certainly did before there were any dams on the stream. On the other hand, huge volumes of water following prolonged winter rain can swell the river to a raging flood. For example, in March 1991, after more than six years of drought, when Gibraltar Reservoir had gone dry and Lake Cachuma had shrunk to less than 20 percent of capacity, enough rain fell in a few weeks to fill all three reservoirs. Ironically, Bradbury Dam, at Lake Cachuma was then in the midst of an earthquake retrofit, and the reservoir could not be filled to capacity, so water that normally would have been impounded was allowed to flow downstream to the sea.

Recently, there has been litigation seeking to restore a perennial flow downstream from Bradbury Dam to the sea in order to provide a suitable habitat for steelhead trout. Irrespective of the ultimate outcome of the litigation, it demonstrates the validity of Mark Twain's famous remark about water in the west: "Whiskey is for drinking and Water is for fighting over." Litigation over the river's flow has a long history going back to at least 1920 when the City of Santa Barbara began to divert water at Gibraltar Dam, arousing the wrath of downstream users.

Los Alamos Valley

The next valley to the north is the Los Alamos Valley, which has developed on a canoe-shaped fold (syncline) in the rocks there. This is a much narrower replica of the Santa Ynez Valley and is drained by San Antonio Creek which enters the sea on the Vandenberg Air Force Base about half way between Purisima Point and Point Sal (Figure 1). Both the Santa Ynez and Los Alamos valleys are bounded on the northeast by a broad band of subdued hills that form an apron below the higher ridges of the San Rafael Mountains. These hills are dissected by a number of stream valleys, some more than 100 feet deep and often with wide, flat floors. The sheet of material that makes up these hills is a gravelly sand with abundant white chips from the Monterey Formation. This gravelly deposit is called the Paso Robles Formation and is a product of streams mostly draining the San Rafael Mountains.

Sisquoc River Valley

Next to the north is the Sisquoc River, one of the county's three longest streams. It extends about 55 miles from its headwaters near Big Pine Mountain to its junction with the Cuyama River at Fugler Point near the hamlet of Garey (Figure 1). Upstream from its junction with Foxen Creek, the Sisquoc River occupies a fairly narrow valley and is generally a perennial stream. Below Foxen Canyon, the valley widens and merges with the Santa Maria Valley. The Sisquoc is the longest undammed stream in the county.

Santa Maria and Cuyama Valleys

Longest of the county's water courses is the Cuyama River which has its headwaters in Ventura County from whence it flows about 100 miles to its junction with the Sisquoc River. Below this junction, the river is known as the Santa Maria River and reaches the sea about 4.5 miles west of the city of Guadalupe.

The fertile Santa Maria Valley is bounded on the south by the Casmalia and Solomon hills and is the widest valley in the county. The western part of the valley extends north into San Luis Obispo County.

Immediately upstream from its junction with the Sisquoc River, the Cuyama River occupies a narrow gorge, the site of Twitchell Dam. In contrast, the lower Sisquoc and Santa Maria valleys merge imperceptibly.

Generally, the Santa Maria River has little flow, not only because its longest tributary, the Cuyama River, is dammed just above its junction with the Sisquoc River, but because of the strongly seasonal rainfall and also because flows that might otherwise reach the Santa Maria River from the Sisquoc River have been reduced or eliminated by withdrawal of groundwater for irrigation and domestic uses. As a result, water flows under the U.S. Highway 101 bridge only after heavy winter rains have saturated the Sisquoc River drainage basin, or when Cuyama River water is released from Twitchell Dam.

The northernmost valley in Santa Barbara County, shared with San Luis Obispo and Ventura counties, is the Cuyama River Valley in the northern and northeastern part of the county. It drains Santa Barbara County's only real desert, an area that lies on the lee side of the Coast Ranges which reduces the effects of winter rainstorms from the Pacific. Much of the upper Cuyama River Valley typically gets less than 10 inches of rain annually and often as little as five inches. It is as dry as much of the better-known Mojave Desert in eastern California.

Because most of the headwaters area is arid, the river is an intermittent stream, albeit quite a long one. The flat parts of the Cuyama River Valley, like other stream valleys in the county, are floored by young flood plain sands and gravels. An appreciable part of the headwaters area, in Ventura County particularly, but also south and east of the twin towns of Cuyama and New Cuyama, in Santa Barbara County, is a typical badland. Badlands are intensely gullied areas and develop under arid and semi-arid conditions where the rocks are soft, weakly consolidated and somewhat impermeable. This combination of qualities inhibits growth of vegetation, and exposes rocks to quick erosion every time rain falls. The rapidity of erosion further limits vegetation by sweeping away many seedlings before they can get established.

The rocks underlying the badland area are mostly non-marine sandstones and conglomerates of Pleistocene age, chiefly old stream and flood plain deposits (Figure 11).

The Cuyama River, like the Santa Ynez River, has somehow managed to cut a narrow, winding gorge across the Coast Range mountains that lie athwart the river's access to the sea. This channel, remarkably, has been cut through some of the more resistant rocks in the county such as the well-cemented Cretaceous sandstones and the Obispo Tuff.

Most geology is hidden underground. When an exploratory well is drilled, geologists can learn a great deal about the hidden past by studying the sequence of rocks recovered as part of the drilling process. A well near the

Figure 11. Cuyama River Valley. Badland topography developed on the marine Miocene Vaqueros and Branch Canyon formations, in the arid, upper part of the valley.

mouth of the Santa Maria River revealed that during the period of low sea level in the last glacial stage of the Pleistocene, about 18,000 years ago, the Santa Maria River cut a gorge into bedrock some 230 feet below present sea level. With the post-glacial rise in sea level called the **Flandrian Transgression**, this gorge was back-filled with river and beach sediments, so that apart from a shallow lagoon at the river mouth, there is no longer any surface expression of this former deep channel. It is likely that a number of other back-filled gorges occur where Santa Barbara County streams reach the sea. The only other such channel known to the writer lies off the mouth of Goleta Slough on the south coast where the seven short streams now emptying into the slough once joined to cut a single bedrock gorge about 200 feet deep, subsequently back-filled with sediment.

Other Stream Valleys and Canyons

Geologists interested in the evolution of landscape are always alert to unusual stream courses because these often provide information on the history of faulting and uplift.

Gaviota Creek is the only stream that cuts right through the Santa Ynez Range. It created the pass where U.S. Highway 101 turns inland from the

Figure 12. Gaviota Pass, looking north. Marine Gaviota sandstones of late Eocene age. The grassy hill in the middle distance is on the Eocene Sacate Formation. The brush covered high ridge in the distance is on the Gaviota Formation.

coast (Figure 12). Apparently, the stream existed before the mountains were uplifted to their present height, and as the mountains rose, the creek stayed in its bed, down-cutting through the rock.

In the Santa Barbara city area, there are several other streams that have unusual courses. The smallest and westernmost of these is a minor, unnamed, intermittent creek that empties into the sea at Hope Ranch Beach. The present stream carries too little water to account for the prominent canyon it occupies. This canyon extends from the Laguna Blanca Basin to the sea and was almost certainly cut by a much larger stream, possibly Arroyo Burro Creek or Cieneguitas Creek. As the hills of Hope Ranch were elevated along the More Ranch fault, the larger stream was diverted to the east if it was Arroyo Burro Creek, or to the west if the culprit was Cieneguitas Creek. However the diversion occurred, it denied the stream from Laguna Blanca to Hope Ranch Beach its former headwaters, leaving a tiny, headless stream too small to have created the gorge it now occupies.

The Lavigia fault crosses this small stream close to where the prominent gorge begins, just seaward of Laguna Blanca. At some point in time, after the larger stream, either Arroyo Burro Creek or Cieneguitas Creek cut this gorge, and was deflected by uplift along the More Ranch fault, the tiny stream left behind could not maintain its course as the block on the seaward side of the Lavigia fault continued to rise. This resulted in a low dam behind which the shallow, undrained Laguna Blanca Basin formed (Figure 9).

Just to the east, Arroyo Burro Creek has a curious course. It and its main tributary, San Roque Creek, rise in the Santa Ynez Mountains, flow across gently sloping land toward the hills of Hope Ranch, abruptly swing easterly toward Las Positas Park, and follow a prominent, narrow valley to the sea at Arroyo Burro Beach. Uplift of the Hope Ranch hills was slow enough in this case so that Arroyo Burro Creek was able to maintain its course across the rising barrier. We call any stream that maintains a course across a rising barrier, an **antecedent stream** because it was there before the barrier was raised (Figure 13).

Similarly, a gorge that crosses a mountain barrier, still occupied by the stream that cut it, is called a **water gap**. Gaviota Pass is a good example of a water gap. On the other hand, where a gorge is no longer occupied by the stream that cut it, the feature is called a **wind gap**. The gorge leading to Hope Ranch Beach is a wind gap.

The present course of Mission Creek through the City of Santa Barbara shows that it has several times been diverted to the west by a slowly rising fold in the vicinity of the Old Mission. Among these older abandoned courses is the one quite easily seen at Santa Barbara High School, which can be traced seaward across the city, though its former path has certainly been somewhat obscured by city streets and buildings (Figure 14).

Sycamore Creek, on the eastern margin of the City of Santa Barbara, is still another example of an antecedent stream that has succeeded in maintaining its course across the rising Riviera-Eucalyptus Hill uplift.

If nothing gets in its way, a stream will follow as straight a course as possible, downhill from its headwaters directly to the ocean. But we can see that many present-day streams in Santa Barbara have been forced to bend away from their original course or even to make right-angle turns around hills being pushed up along faults. This is very good evidence that uplift, folding, and faulting of the underlying rocks are going on right now in Santa Barbara. Unless we happen to see abrupt uplift accompanying an earthquake, most of us are unlikely to be aware that the earth is being deformed beneath our feet. It's akin to seeing the movement of the hour hand on a clock over the course of a few seconds.

Figure 13. Antecedent valley of Arroyo Burro Creek near Santa Barbara. The cliffs and grassy hills in the foreground are developed on marine Monterey shale. The grassy hills in the middle distance are on the Rincon Formation. The Lavigia fault lies on the far side of this grassy hill and crosses the road near the prominent bend. Veronica Springs are located just to the left of Arroyo Burro Creek near the middle of the photo.

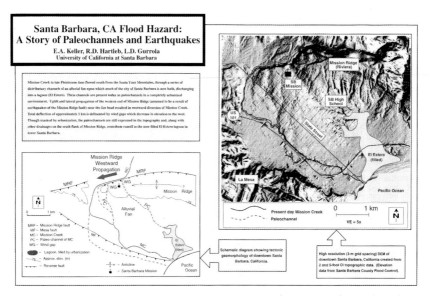

Figure 14. Santa Barbara city showing the growth of Mission Ridge. (Courtesy of E. A. Keller, R. D. Hartleb & L. D. Gurrola).

LAKES

In the normal course of geological events, lakes are fairly temporary features because streams tend to fill them with sediment and at the same time drain them by eroding their outlets.

Where rocks are resistant to erosion or the landscape is subdued and covered with dense vegetation, lakes may persist for quite a long time. Such conditions are not characteristic of our part of California and as a result, there are very few natural lakes in the southern Coast Ranges.

Apart from some small coastal lagoons and small bodies of water at mouths of streams, Santa Barbara County has only one permanent, natural lake, Little Zaca Lake near the geographic center of the county (Figure 15). The lake lies at an elevation of 2396 feet in a bowl-shaped valley formed by a pair of landslides that slid northward from Zaca Ridge, perhaps 10-12,000 years ago, blocking the headwaters of Zaca Creek and impounding not only Zaca Lake itself, but a small, intermittent pond just west as well. The rocks of the mountains around the lake belong to the marine Monterey Formation of middle Miocene age and show some dramatic anticlinal folding on the north side of the lake.

Figure 15. Zaca Lake. Note the tightly folded strata of the marine Miocene Monterey Formation.

There are three man-made water supply lakes on the Santa Ynez River. The uppermost one was dammed in 1930 by Juncal Dam forming Jameson Lake, from which water is delivered to Montecito via the Doulton Tunnel. The Doulton Tunnel acts as a long horizontal well and produces 500 to 700 acre feet of water annually in addition to the water it delivers from Jameson Lake.

Downstream from Jameson Lake is Gibraltar Reservoir, built by the City of Santa Barbara in 1920. Water from this reservoir is brought through Mission Tunnel to Sheffield Reservoir located in the foothills of Santa Barbara. Like the Doulton Tunnel, the Mission Tunnel acts like a horizontal well, yielding an average of 1100 acre feet of water annually in addition to that delivered from Gibraltar Reservoir. Interestingly, the Mission Tunnel was completed in 1912, well before Gibraltar Dam was finished. During the interim, groundwater seeping into the tunnel furnished a significant share of the city's water needs.

Lake Cachuma, impounded by Bradbury Dam in 1953, is the largest and most recent of the three reservoirs. Water is carried through the 7-mile long Tecolote Tunnel to Glen Annie Reservoir on the south coast. Like the other two tunnels, the Tecolote Tunnel yields an average of 2000 acre feet of water each year.

The only other significant man-made lake is Twitchell Reservoir on the lower Cuyama River near Santa Maria. This reservoir was built by the U.S. Bureau of Reclamation in 1958 for flood control and for recharging the Santa

Maria Valley groundwater basin. It protects the flat Santa Maria Valley from periodic high-intensity flooding originating in the Cuyama River drainage. Because the drainage basin is quite arid with sparse vegetation, high-intensity rains are likely to produce rapid runoff and high sediment loads. As a result, each flood episode can be expected to further reduce the storage capacity of the reservoir.

Laguna Blanca in the Hope Ranch area west of Santa Barbara is, in part, a natural lake. This small lake occupies a shallow depression between the More Ranch and Lavigia faults and was filled only intermittently before Hope Ranch was developed in 1925. At that time, a well in the foothills east of San Marcos Pass Road fed a pipeline to the lake, providing a permanent supply. Deterioration of this pipeline caused it to be abandoned in the 1960s and municipal water supply was used to augment the lake when winter rains failed to provide sufficient water. As water supplies have become more restricted and expensive, the lake has been allowed to go dry during severe drought.

SPRINGS

Warm or hot springs in the county are generally associated with faults, such as Las Cruces (Gaviota) Hot Spring, because faults allow water to move upward from heated rocks at depth. Cold or mineral springs may or may not be fault related. There are no springs associated with volcanic activity because there are no young volcanic rocks anywhere in the county. (The youngest volcanic rocks are middle Miocene in age, or about 15 million years old.)

There are a number of springs, warm and cold, in various parts of the county. Perhaps the most famous of these, at least in a commercial sense, is Veronica Springs in Arroyo Burro Canyon west of Santa Barbara. This spring produces water with an appreciable content of epsom salt (hydrous magnesium sulfate), a bitter but very effective laxative. Water from this spring was bottled and marketed for many years beginning about the turn of the century and ending about 1960. The springs still exist, but the water is no longer bottled. The water comes out of the Monterey Formation, the source of the epsom salt.

Several small hot springs with temperatures of about 116°F occur in Hot Springs Canyon near Montecito. These springs issue from the lower part of the Eocene Cozy Dell Formation and have been used for bathing since about 1860. Various health spas and even a hotel utilized these springs from the late 19th century until about 1960, when the last buildings were destroyed by a forest fire.

COASTAL PLAINS, BEACHES
AND ELEVATED MARINE TERRACES

One often hears the expression, "The Goleta Valley" or "The Carpinteria Valley", but these are not strictly valleys; they are narrow coastal plains, not exceeding five miles in width anywhere on the south coast.

The coastal plains in southern Santa Barbara County have complex origins. Some are partly benches or platforms cut by wave erosion (marine terraces), but many are also partly flood plains formed by the various streams draining the south face of the Santa Ynez Mountains. And still others are in part the result of faulting.

At Carpinteria and Goleta, extensive salt marshes occupy parts of the coastal plain. The origin of these features will be discussed more fully later in this section; it will suffice for now to point out that these marshes are geologically transitory or ephemeral features that will disappear as erosional and depositional processes adjust to the post-Glacial rise in sea level that occurred about 6000-8000 years ago.

Because most of the county's shoreline is undergoing geologically rapid uplift, cliffs and narrow beaches are usual. Only where major streams have widened their valleys at the shore, or where local subsidence of the land has taken place, do we find low-lying coasts.

On the more exposed, west-facing coast north of Point Arguello, as far as the Santa Maria River, cliffs and narrow beaches are the rule except at stream mouths. What distinguishes this coast from the more sheltered south-facing coast, are its large sand dunes. Strong northwesterly winds rake this shoreline throughout the year and drive beach sand inland to form widespread dunes that extend inshore several miles at some places.

Under undisturbed, natural conditions, beaches that supply this sand are themselves regularly replenished by sand contributed to the coasts by various streams and rivers. However, when dams are built, some supply is withheld from the beaches and eventually the beaches will become narrower and contribute less new sand to coastal dunes. Further, as a beach narrows, it affords less and less protection from wave erosion of cliffs or coastal structures.

Northwesterly winds create waves that approach the shore obliquely. The waves wash up the beach at an angle, but return to the sea perpendicular to the coast (Figure 16). As a result, both the water and any sand entrained in it, follow a zig-zag path along the coast. For Santa Barbara County's west-facing coast, this flow, called the **longshore current**, moves southerly. On the south coast of the county, most frequent winds and waves approach from the west and the current is easterly. Some beach sand grains are rolled along the surface and do not get carried upward into the water. Collectively, these sand grains move in the direction of the longshore current by a process called **beach drift**. Sand may be transported either by the longshore current or beach drift depending upon grain size and strength of waves.

On Santa Barbara County's west-facing beaches, sand is supplied by streams in southern San Luis Obispo County, the Santa Maria and Santa Ynez rivers and by San Antonio Creek. Dams on some of these streams will surely reduce the coastal sand supply in the coming years.

One may well ask what becomes of these coastal streams of sand because under natural conditions, the beaches seem stable over the long term. On the west-facing coast, a lot of this sand is carried inshore by the wind, but there are few dunes on the south coast and the beaches there and in Ventura County seem fairly stable over the long term. We now know that most beaches can be divided into "cells" in which sand supply and sand loss are in rough balance. For southern Santa Barbara County, the cell begins at Point Conception on the west and extends east and south to Point Hueneme in Ventura County. Sand is supplied almost entirely by streams entering the ocean in this cell. Santa Barbara south coast streams are short and steep and make relatively modest sand accumulations, but the Ventura and Santa Clara rivers in Ventura County make major contributions and as a result, sand supply south of Ventura is about three times as much as it is in Santa Barbara County (Figure 17).

At Point Hueneme, a deep sea-floor or submarine canyon heads just off the harbor entrance. This canyon is the sluiceway draining sand from the beach into very deep water offshore. The consensus today is that such canyons owe their origin and maintenance to such streams of sand.

As for the west-facing beaches north of Point Arguello, aerial photography shows that a submarine stream of sand flows southerly off this point entering a canyon at that headland. This canyon was once thought to head some distance offshore, but modern surveys show that it begins close to the beach.

The west-facing beach cell probably begins near Pismo Beach in San Luis Obispo County and ends at Point Arguello. Because of the orientation of this shoreline and its greater windiness, a good deal more beach sand is lost to coastal dunes than is the case for the south coast cell.

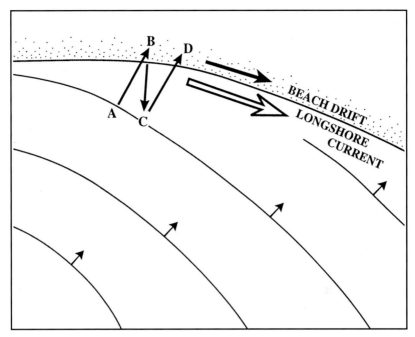

Figure 16. Operation of the longshore coastal current. Waves approaching at an angle (A to B and C to D); backwash is from B to C. This results in zig-zag transport along the coast. Beach sand drifts down coast in the same manner.

Beaches are not only responsive to total sand supply but react to seasonal changes that may be quite dramatic. As a rule, California beaches are flattened and lowered during the stormier winter months when closely-spaced or "short period"[1] waves strip sand from the beach and move it out into shallow water, often exposing the rocky, wave-cut platforms that are concealed beneath the sand most of the rest of the year. As summer comes on, and the North Pacific storms abate, the short-period waves give way to long-period or widely spaced waves travelling from winter storms in the Southern Hemisphere. These waves drive the sand shoreward, cover exposed rocks and make the beaches somewhat steeper.

Studies done at Santa Barbara by the U.S. Army Corps of Engineers about 1940 showed that the volume of sand moving along shore was also

---

[1] Period is the time interval between waves passing a point. Short-period waves have intervals from about 7-10 seconds; long-period waves as much as 20 seconds.]

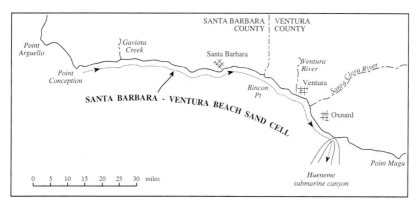

Figure 17. The Santa Barbara-Ventura beach sand cell. The volume in transit increases easterly until it is shunted into deep water via the Hueneme submarine canyon.

affected by the rainfall pattern in the previous 3.5 years. In other words, a wet winter with lots of stream runoff would be expected, 3.5 years later, to cause an increased flux of sand along the coast resulting in wider beaches. This happens because heavy stream runoff events deposit delta-like accumulations of sand at stream mouths which are slowly removed by the longshore current and transported down the coast (Figure 16).

The relationship between total annual rainfall and volume of sand moved alongshore is imperfect. This results because total rainfall may come in intense storms close together producing floods that carry a lot of sediment to the beach, or by many gentler storms more widely spaced which may not produce any flooding and therefore transport only minor amounts of sediment to the beach.

Because south coast beaches are dependent upon sand from the short, steep streams that drain the Santa Ynez Mountains, the sand supply is limited. Despite this, the Army Corps' study showed that the volume of sand entering Santa Barbara Harbor varied from several hundred cubic yards per day in the quieter summer months to several thousand yards per day during winter stormy periods.

Santa Barbara Harbor was one of the first places in the country where the volume of sand carried by the longshore current was quantified, and where problems associated with breakwater construction on coasts with no natural harbors was first appreciated and understood. What had not been understood until then, was that sand transport and the longshore current are maintained by breaking waves approaching the shore at an angle. Breakwaters are built to protect the enclosed area from breaking waves, and without breaking waves, the current slows to a stop and drops whatever sand it carries.

Before 1928, Santa Barbara had no harbor and visiting ships had to anchor offshore and transport passengers and cargo through the surf, or after 1872 make use of Stearns Wharf when weather permitted. Beginning in 1873, local citizens, on five separate occasions, petitioned the Federal Government for advice and assistance in constructing a breakwater to create an artificial harbor. The Army Corps of Engineers reviewed each of these requests and recommended each time that no breakwater be built. The last of these requests was in 1921. Despite this, local interests prevailed and $750,000 was raised in 1928 to construct an L-shaped breakwater west of Stearns Wharf. The long arm of the L was parallel to the beach and the short arm extended toward, but did not reach Point Castillo. It was evidently supposed that by leaving a gap near shore, sand transported along the coast would pass uninterruptedly through the harbor (Figures 18 and 19).

It became very evident during 1929 and 1930 that sand entering the harbor through the gap would soon shoal the harbor and greatly reduce its usefulness. The solution seemed obvious: Complete the short breakwater leg to Point Castillo and stop the influx of sand through the gap. This was done and it worked—for a while.

The sand being moved eastward by the longshore current had to go somewhere, so it was deposited where the current stopped—at the west or up-current side of the new breakwater leg. In about four years, the current had filled a triangular area from the beach to the outer leg of the breakwater and then began to move sand along the outer face of the breakwater into the harbor entrance (Figure 21). During this period, the city gained a sizeable chunk of land, the site of the present harbor business area, the Yacht Club, La Playa field, and Leadbetter Beach. On the other hand, beaches east of Santa Barbara as far as Rincon Point were starved because their normal sand supply was being trapped in and around Santa Barbara Harbor, and the natural erosional processes continued unabated to the east. Serious beach erosion occurred as far east as Carpinteria.

By 1935, sand had nearly blocked the entrance to Santa Barbara Harbor, and a hopper dredge was brought in to clear the entrance. A hopper dredge is a scow with large doors in the bottom of the hull. It is unloaded by opening these doors, but that requires water deep enough to permit the doors to be operated. As a result, the dredged sand was dumped in about 18 feet of water

Figure 18. City of Santa Barbara in 1928 during construction of the breakwater for Santa Barbara Harbor (lower middle); vertical view. (Photo by Fairchild Aerial Surveys).

off East Beach, forming a submarine ridge nearly 1000 feet long. In the 60 or so years since, this submarine ridge is still pretty much in place because it lies outside the surf zone where longshore currents are generated. As a result, downcast beaches continued to endure erosion until the harbor was dredged again in 1938. This time, a suction dredge was used and sand was pumped onto the beach east of Stearns Wharf. Beaches to the east quickly began to recover and by 1942, beaches about three miles east of Santa Barbara had been restored to about their condition in 1929. However, beaches farther east were still showing severe erosion until about 1940. At Santa Claus, just west of Carpinteria, a strip of beach 240 feet wide eroded during one especially stormy winter, destroying several buildings and parts of buildings, resulting in property losses of about $2 million.

Suction dredging has continued intermittently to the present and the beaches east of Santa Barbara have been stabilized so far as sand transport is concerned, though they are of course subject to seasonal and other changes induced by the weather.

Figure 19.   Santa Barbara and the harbor area in October 1929. Note the bulge in sand (arrow) transported through the gap between the breakwater and Point Castillo before it was closed. (Photo by Fairchild Aerial Surveys).

Unfortunately, these lessons were seldom heeded, even in California, and there are many examples of costly mistakes that could have been avoided by referring to the history of Santa Barbara Harbor (Figures 18, 19, 20 and 21).

Coastal Erosion at Santa Barbara & Goleta

Seacliffs dominate the coast from Rincon Point to Point Conception and beyond, and all of this cliffed coast is subject to erosional retreat. In a sense, it is as if nature abhors cliffs of any kind and seeks to erase and flatten them as quickly as possible. The rate at which cliff destruction takes place depends not only on the resistance of the rocks involved, but on the mix and intensity of destructive processes as well.

As far as seacliff erosion in Santa Barbara County is concerned, most cliffs are composed of relatively soft, easily eroded sedimentary rocks. Harder rocks occur at a few places on the west-facing coast north of Point Arguello and at many places on the Channel Islands, but nowhere in the county are there any cliff-forming rocks as resistant as the hard, fresh granitic rock exposed in such places as the high walls of Yosemite Valley.

Figure 20. Santa Barbara Harbor area in December 1934. Following closure of the gap from the breakwater to Point Castillo, sand deposition rapidly occurred between the breakwater and Santa Barbara Point, forming Leadbetter Beach. The Santa Barbara Mesa Oil Field can be seen on the uplifted terrace west of the harbor (upper left or arrow). (Photo by Spence Air Photos).

Where the sea regularly reaches the base of the cliffs, marine erosion can be quite vigorous. Wave energy first loosens rocks at the cliff base and adds this debris to any brought to the beach by stream floods or to any rocks that fall from the cliff above the reach of waves. Wave action uses these rocks as an abrasive to cut a notch at the base of the cliff. As this process continues, the notch is enlarged and deepened until the rocks above become unstable and fall to the beach, adding more grist to the horizontal sawing process carried on by waves (Figures 22 and 23).

In many places, however, a protective beach absorbs most of the wave energy and only at the highest tides or during storms do the waves reach the cliff base. As a result, non-marine erosional processes become the dominant factors in cliff erosion in such places. For much of the urbanized area around Santa Barbara and Goleta, seacliff retreat is almost entirely due to non-marine erosion, the debris from which is periodically cleaned away by storm waves or very high tides.

Figure 21. Santa Barbara Harbor in 1958. Note the sand bar (arrow) forming inside the breakwater from sand transported along the outer leg of the breakwater from Leadbetter Beach to the west. (Photographer unknown).

SAND DUNES

In order for a sand dune to develop, there must be a sand source such as a bare desert wash or a beach where either streams or wave action prevent the growth of protective vegetation. Strong winds capable of picking up loose sand are also necessary. Finally, there must be a factor or factors that cause the wind to drop some of its sand load. Among the possible factors are rocky ridges that slow the wind and provide a quiet zone on the lee or downwind side of the obstacle. Shrubby vegetation may likewise provide enough interference to cause some sand to be dropped on its downwind side.

Some sand is always picked up and blown off dune surfaces, so if the dune is to remain stable in size, the volume of incoming sand must match the volume of sand blown away. If the source of sand should be cut off, the dune will gradually shrink and may even disappear. Dunes grow where the volume of arriving new sand exceeds that which is lost.

Interestingly, once a small dune forms, given sufficient incoming sand, it will tend to grow because of the ways in which sand is carried by the wind. When winds are very strong, considerable sand may be carried long distances

Figure 22. Gaviota State Beach; view west. Seaward-dipping strata of cherty shale from the marine Miocene Monterey Formation. Note the prominent wave-cut bench. The darker rocks at the water line are tarry conglomerates.

Figure 23. Santa Barbara Point. The rocks on this wave-cut platform belong to the marine Miocene Monterey Formation.

in **suspension**, that is, it remains aloft. Some grains, often the heavier or larger ones, bounce along by a process called **saltation**. Whenever these grains strike a rock or other hard surface, they leap up into the air and move by suspension, but when they strike a soft, yielding surface like that of a dune, they do not bounce, but are added to the dune. Finally, if the grains are too large or heavy to be moved by a wind of a given strength, they may just be rolled along the surface.

Dry sand cannot be piled up more steeply than about 33°. Even if the wind isn't blowing, the position of the slip-face will show the direction of the last sand-moving wind.

On Santa Barbara's west-facing coast, especially near the mouth of the Santa Maria River, is an extensive tract of dunes, extending inland 20 miles in some places (Figures 24 and 25). South of Point Sal, dunes extend inland almost five miles from the beach. Narrower dunes lie inshore of Purisima Point and near the mouth of the Santa Ynez River. There are very few dunes along the south-facing coast because westerly and northwesterly winds tend either to blow the sand along the beach, or offshore into the sea. However, there is one small area of dunes near Goleta at Coal Oil Point where a short stretch of west-facing beach occurs.

Figure 24. Guadalupe Dunes near the mouth of the Santa Maria River; view south. The Casmalia Hills in the distance are mostly made up of marine Miocene Monterey Formation with lesser amounts of older rocks.

Some minor dune fields occur on the western and northern coasts of Santa Rosa and Santa Cruz islands, but the most prominent dune field on any offshore island, apart from San Nicolas Island in Ventura County, is the on San Miguel Island (Figure 26). About 30% of this island is covered by dune sand mostly driven inshore from Simonton Cove on the north coast. This dune tract extends half way across the island and much of it is carried into Cuyler Harbor on the northeast side.

Some of the San Miguel dunes are characterized by what have been called "Ghost trees", limy, tubular structures projecting out of the sand (Figure 27). These appear to be calcified casts of plant roots and stems, exposed by erosion of the dunes following destruction of the formerly more extensive vegetation. It is not entirely clear whether this erosion is part of a natural process or due in some measure to overgrazing by sheep. An identical situation occurs on San Nicolas Island.

The most extensive tract of dunes in the county lies south of the Santa Maria River mouth and is known as the Guadalupe Dunes. The corresponding dune field north of the river mouth in San Luis Obispo County is known as the Nipomo Dunes. Many years ago, geologists recognized that the Guadalupe Dunes consisted of three distinct sand sheets (Figure 25).

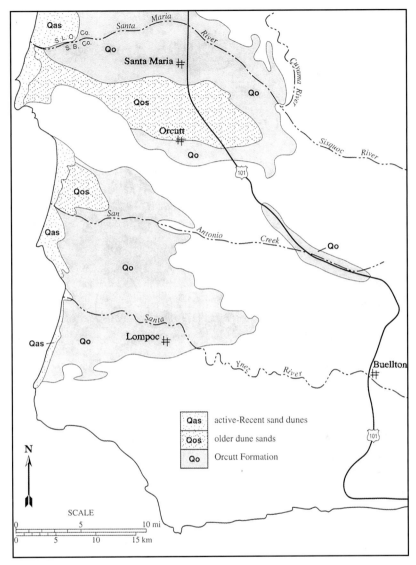

Figure 25. Diagram of western Santa Barbara County showing the distribution of the three ages of dune sand. (Courtesy of H. E. Ehrenspeck).

The oldest unit, of Pleistocene age, is known as the Orcutt sand and generally extends farthest inland. In some places this sand blanket reaches an elevation of 1100 feet above sea level. It should not be concluded that the Orcutt sand is 1100 feet thick; like the younger sand sheets, the Orcutt sand is a much thinner blanket covering older rocks.

Figure 26. Simonton Cove, San Miguel Island in 1988. Note the streams of beach sand blown across the island toward Cuyler Harbor. (Photo by W. B. Dewey).

The Orcutt sand is usually brownish, due to staining of the sand grains as iron-bearing minerals in the sand are weathered. Most of the Orcutt sand is weakly cemented and has been stabilized by grass or other vegetation so little of it now forms active dunes. Much of this old dune sand is now covered by houses and shopping centers in the Orcutt area, though some exposures can be seen in road cuts along U.S. Highway 101.

The next youngest dunes are found closer to shore, are generally paler in color, and less frequently stabilized by vegetation. They often comprise the higher, inland parts of the dune field, which in some places rise more than 870 feet above sea level. These younger dunes extend inland as much as five miles and overlap the older Orcutt Sand.

Closest to shore are the modern, active dunes currently forming from sand driven inshore from the beaches. These dunes are pale gray or white and are seldom covered with much vegetation. Because of the persistent windiness of the coastal area and the copious sand supply, these dunes occasionally reach elevations of 50 feet or so.

Smaller patches of dune sand lie south of Point Sal near Purisima Point, and south of the mouth of the Santa Ynez River as far as Point Pedernales.

Figure 27. San Miguel Island. The "ghost trees" represent calcified casts of plant stems and roots exposed by erosion of surrounding sand. (Photo by D. L. Johnson).

HEADLANDS AND POINTS

Headlands and points along the coastline are prominent features that indicate a number of special origins and circumstances, and for these reasons are worth special attention. The majority of the county's headlands are rocky cliffs, which are somewhat more resistant to wave attack than adjacent areas. There is more variety than at first may be evident, as the following tour from Rincon Point at the southeastern corner of the county to Rocky Point near the northwestern corner will show. Points and headlands on the offshore islands, where notably distinctive, are considered in the discussion of the various islands.

Rincon Point

This headland is what is called a small **cuspate delta**. It is a triangular deposit with a single, centered stream course whose mouth is at the apex of the triangle. The Rincon Creek delta is formed from boulder deposits carried down to the shore by Rincon Creek during flood events (Figure 28). Rincon is widely regarded as one of the best winter surfing spots in California.

Figure 28. Rincon Creek, Carpenteria area in July 1953; vertical aerial view, Note the small cuspate delta at the creek mouth (arrow). The oval between the highway and the ocean (arrow) is the former Carpinteria Thunderbowl, a small sag pond on the Carpinteria fault. (Photo by of U. S. Department of Agriculture).

Sand Point

This headland is located at the entrance to Carpinteria Salt Marsh (El Estero) and has quite a different origin (Figure 29). It has developed because just offshore is a rocky reef exposed only a few times each year during extremely low spring tides. (Incidentally, the term **spring tide** does not refer to the season, but rather to the lowest tides each month near the times of full and new moon.)

The rocky reef at Sand Point is composed of resistant beds of the Monterey Formation raised along the Rincon Creek-Carpinteria fault. It acts as a natural breakwater, reducing waves on its inshore side. This has slowed the longshore current and caused the beach to build outward toward the barrier. Eventually it enclosed a coastal lagoon which was the predecessor of the present salt marsh. Beaches of this sort are called **sand spits** and if they eventually tie the offshore reef or island to the mainland, they are called **tombolos**. Sand point is not yet a tombolo, though Point Sur and Point Reyes on the northern California coast are good examples.

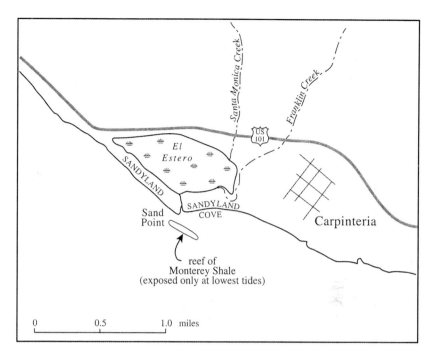

Figure 29. Map of the Carpinteria area showing development of the bulge in the coastline (an incipient tombolo) at Sand Point.

The exclusive residential areas known as Sandyland and Sandyland Coves are on sand spits, and as such are vulnerable to severe erosion during large storms.

Salt marshes like Carpinteria are, from a geologic point, very transitory features and represent an intermediate stage between an open lagoon and dry land. Streams draining into the salt marsh will eventually fill it with sediment.

Loon Point

This point is at the eastern end of Summerland Beach and exists chiefly because folding and faulting (supplemented by gravels from Toro Creek) have raised the soft rocks of the Casitas Formation rapidly enough so that wave erosion has not yet had time to erase this minor headland (Figure 30).

Figure 30. Loon Point near Summerland. The anticline in the tightly folded, nonmarine middle Pleistocene Casitas Formation is dimly visible in the highest part of the cliff (arrow). Folding of these very young rocks demonstrates recent tectonic activity.

Fernald Point

Fernald Point is at the western end of Summerland Beach and is easily seen from U.S. Highway 101 where it crests Ortega Hill. It is a minor, but quite scenic feature favored by the surfing community because of its often distinctive wave pattern. It is mainly the result of a gravel delta deposited by Romero and San Ysidro creeks.

Point Castillo

This was once a prominent headland where the present Santa Barbara Breakwater connects with the shore. It is no longer very evident because of harbor and road development as well as its use for rip-rap for shoreline protection along Leadbetter Beach. It also once provided a very fossiliferous exposure of the Plio-Pleistocene Santa Barbara Formation and was known in the older literature as the Bath House Beach locality (Figure 20). The Santa Barbara Formation at Point Castillo is slightly more resistant to erosion than the Rincon Formation present to the west, and this, plus some uplift on the Mesa fault accounts for this now departed headland.

Santa Barbara Point

This headland is formed from the Monterey Formation and is located at the western end of Leadbetter Beach. It marks the location of a small, tightly-folded anticline and is partly due to the presence of the less resistant Rincon Formation to the east that has allowed wave erosion to cut more deeply into the coastline east of Santa Barbara Point.

More Mesa Tar Deposit

At the western end of More Mesa Beach, near Goleta, is a massive, inactive tar seep that has formed a resistant, bulbous mass that resists wave attack (Figure 31). This tar seep has incorporated beach sand and some rock chips and forms a whale-shaped mass composed of a sort of natural macadam. Some minor, more recently active tar seeps issue from the Santa Barbara Formation exposed in the 80-foot cliff nearby. Just offshore are several other tar seeps.

Goleta or Campus Point

This is the next prominent headland and is located on the UCSB campus. Rocks here are weak marine mudstones of the late Miocene Sisquoc Formation that usually yield quickly to weathering and erosion. Here, however, Sisquoc rocks are slightly more resistant than the rocks either immediately to the east or west, possibly because part of the headland once had a protective coating of tar from a nearby offshore seep. In addition, the UCSB campus and Isla Vista are on a block that was raised 30-40 feet above sea level on the More Ranch fault. This uplift almost certainly was due to a number of separate events that raised this old platform of marine erosion to its present height. The initial uplift must have been on the order of 10 feet because it trapped many rock-boring clams in their burrows where they remain to this day. They can easily be seen in the cliff face between Goleta and Coal Oil points. Dating has shown these clams to be about 40,000 years old.

This tale is complicated because we also know that sea level dropped world-wide during this same time period, as the large ice-age glaciers spread over northern land areas. We also know that sea level today is about as high as it has been in the geologically recent past, so we cannot say that the wave-cut platform is due simply to a drop in sea level. Furthermore, sea level changes take place relatively slowly, so are very unlikely to trap large numbers of rock-boring clams in their burrows as actually occurred. Terrace elevation, therefore, had to be due to uplift.

Figure 31. More Mesa Beach near Goleta. Tar seep emerging from the marine Plio-Pleistocene Santa Barbara Formation.

In the immediate vicinity of Goleta Point, however, there is evidence of both abrupt uplift and a drop in sea level. The ice-age drop in sea level is demonstrated most strikingly by an oil company discovery about fifty years ago, when a bedrock channel, now completely filled with sediment, was found off the mouth of Goleta Slough. This buried channel is at least 200 feet deep, matching rather well the known fall in the ice-age sea level. It appears to have been cut by a stream whose six or seven tributaries now enter Goleta Slough separately. These tributaries evidently joined to cut a channel to the old shoreline which then lay some distance seaward of the present shore. As sea level returned to about its present elevation somewhere between 17,000 and 6,000 years ago, this old channel and the lower parts of its six or seven tributaries were flooded forming a coastal lagoon where Goleta Slough is now located. Most of this former lagoon has now been converted to dry land by natural stream deposition, the growth of marsh vegetation and most signifcantly, by filling a good bit of the area to form Santa Barbara Airport.

The lagoon on the UCSB campus is part of an old drainage system now completely cut off from what was very likely a connection with the Goleta Slough ice-age stream system. We can only guess what the connection looked like or where it was because erosion of the sea cliffs has erased much of the evidence (Figures 32 and 33).

Prior to about 1960, the campus lagoon was a dry salt flat most of the year, filling with sea water only when the highest tides washed over the two low sandy barriers that separate the lagoon from the sea. When the UCSB campus marine laboratory was built, it was decided to raise the two barriers slightly and discharge sea water from the laboratory into the lagoon in order to keep it filled. This certainly made it more attractive esthetically and also eliminated the unpleasant fragrance that always occurred during the final stages of drying.

Coal Oil Point

Coal Oil Point is located at the mouth of Devereux Slough just west of Goleta Point (Figure 33). The origin of this headland is similar to Goleta Point and Slough to the east and involves disruption of an ice-age drainage system by uplift on the More Ranch fault and subsequent marine erosion (Figure 2). The name of the point comes from the strong odor of crude oil, especially noticeable on calm days when a pronounced oil slick develops over the large submarine oil and gas seep just offshore. Some years ago, an oil company, in

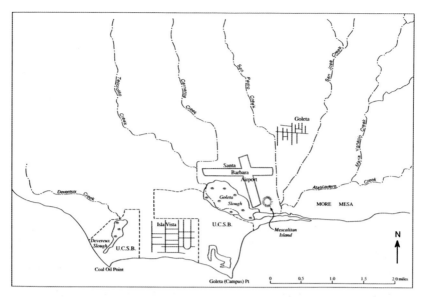

Figure 32. Map of the Goleta Slough, UCSB campus, and vicinity showing drainage patterns.

an environmental trade-off, installed an underwater "umbrella" over the seep to catch the escaping gas and oil in order to reduce air pollution and tar on the beaches. Somewhat to everyone's surprise, in addition to the expected five or so barrels of oil daily, this device trapped about 500,000 cubic feet of gas each day. This seep is only one of many offshore seeps in the Santa Barbara Channel whose tarry emissions have long plagued beach-goers on the south coast. One of the persistent southern California myths is that the tar on the beaches comes from oil tankers operating offshore or from drilling platforms.

George Vancouver, when he visited the area in 1792, and Pedro Font, in 1776, both observed oil slicks on the sea and tar balls on the beaches, long before there were oil tankers or offshore drilling. Further, the Chumash Indians living in the area used the tar for caulking their boats, probably for centuries.

A short stretch of west-facing beach occurs at Coal Oil Point, the only example on the south coast of Santa Barbara County. It is here that the only coastal sand dunes on the south coast occur because the prevailing westerly winds drive beach sand inshore rather than parallel to the beach.

Chapter 2 55

Figure 33. Goleta Point with the lagoon on the UCSB campus in the foreground and Coal Oil Point in the distance. All the sea cliffs in this area are cut in the marine Miocene Sisquoc Formation. The Coal Oil Point anticline and oil and gas seep lies just offshore from that point.

El Capitan Point

At El Capitan State Beach is a small boulder delta very much like the one described at Rincon Point to the east. Cañada del Capitan Creek has delivered the boulders from the Coldwater and Matilija sandstone exposed near the summit of the Santa Ynez Mountains. The size of the boulders is evidence of large floods and probably mudflows in the past (Figure 34).

The Headland at Refugio Beach

What looks like a headland when seen from Refugio Beach is more the result of a small embayment at the mouth of Refugio Creek. West of the creek mouth, the seacliff is armored by the relatively resistant Monterey Formation, but at the creek mouth the less resistant Rincon Formation is present. This has encouraged marine erosion and helped to form the embayment, aided, of course, by creek erosion.

Furthermore, Refugio Creek ends in a small lagoon at the beach, the result of the post-glacial rise in sea level that flooded the old ice-age creek

Figure 34. El Capitan State Beach; view west. Most of the boulders in the delta are derived from the marine Eocene Coldwater and Matilija formations exposed in the higher parts of the Santa Ynez Mountains. The green and black coatings on the rocks are marine algae, not oil.

channel. There may also be a filled bedrock channel off the present creek mouth, though no evidence of such a feature has yet been found.

Point Conception

This point is perhaps the most famous and certainly one of the more prominent headlands on the California coast (Figure 35). It is located about 48 miles west of Santa Barbara, where the east-west trending shoreline begins its change to a nearly north-south trend beyond Point Arguello.

Point Conception is an oceanographic, climatic and biological boundary. Because of the protection it provides from the prevailing northwesterly wind, marine waters to the south and east are warmer and more sheltered. Many organisms, land and marine, plant and animal, reflect this prominent climatic transition point. Southern types in many cases reach their northern limit near Point Conception, and similarly, northern species may not extend southeast of this point, though some of the northern marine species cross this boundary by occupying deeper, colder waters.

Figure 35. Point Conception. Note the small, anticlinal fold in the marine Miocene Monterey Formation exposed in the headland to the left (arrow).

Point Conception is actually a double point; the southeastern part is called Government Point and the northwestern one, Point Conception. Both are cliffed headlands carved from the Monterey shale by wave action. They are backed on the landward side by a nearly flat, uplifted marine terrace about two miles wide. At Point Conception, however, the ground rise is due to a small anticlinal fold whose axis cuts across the tip of the point; most of the axis of this fold lies offshore. This fold has arched up the Monterey Formation and contributed to the high seacliffs, which rise about 150 feet above the water. The fold can be seen in Figure 35.

The Point Conception lighthouse is a prominent feature on this otherwise lonely coast. It was built shortly after California became a state in 1850. Because this was about 25 years before Thomas Edison invented the electric light, the light source was a kerosene mantle, a large-scale example of the Aladdin storm lamps that can be purchased even today. By focusing this light from a glowing mantle, a beam of 630,000 candle power could be produced. It was not until 1948 that electric power reached this remote bit of coast, allowing the old kerosene beacon to be retired, one of the last of its type in the United States.

Between points Arguello and Conception is a large open embayment facing southwest, cliffed for the most part, but with a stretch of sandy beach and some coastal dunes at the mouth of Jalama Creek.

Figure 36. Point Sal and Lion Rock. Most of the cliffs in this area are cut in the Point Sal ophiolite sequence of Jurassic-Cretaceous age.

Point Arguello

This marks the place where the coastline begins its northerly trend. Like Point Conception, Point Arguello is a complex, rocky headland of folded Monterey Formation capped by a marine terrace whose elevation near shore is about 100 feet, rising inland to about 250 feet. The easternmost part of this headland has no official name, but the other three parts are named, from south to north, Rocky Point, Point Arguello and Point Pedernales. The last is formed from Miocene Tranquillon volcanic rock. These volcanic rocks are exposed at a number of places in the westernmost part of the Santa Ynez Range, especially near Tranquillon Mountain (2170'). These volcanic rocks are mostly lava flows at the inland locations, but at Point Pedernales they are mostly tuffs (ash beds), and agglomerates (blocky volcanic rubble). The other parts of the Point Arguello headland are exposures of Monterey Formation.

Purisima Point

This headland lies about midway between the mouth of the Santa Ynez River on the south, and the mouth of San Antonio Creek on the north. Like many of the south coast headlands, this one is yet another exposure of the Monterey Formation, but is here capped by a considerable expanse of dune

Figure 37. Point Sal, north side. Dr. Gordon Haxel sitting on an example of Point Sal pillow lava. Pillow structure is evidence of the submarine eruption of lava. (Photo by C. A. Hopson).

sand, almost a mile wide at the point, but increasing to a width of nearly four miles north of San Antonio Creek.

Point Sal

This point is surely the most interesting geologically of all the headlands in the county because of its unique set of rocks (Figure 36). Point Sal is the western or seaward end of the prominent Point Sal Ridge whose highest point is Mount Lospe (1640'). Although the Point Sal Ridge includes a variety of rock ages and types, it is of special interest because of the presence of an unusual suite of rocks called an ophiolite. This sequence is composed of deep sea sedimentary and volcanic rocks added to the continental margin during the process of subduction. It will be recalled that subduction refers to the events that take place when an oceanic crustal plate is pulled or pushed beneath an adjoining continental plate as sea floor spreading adds to the oceanic plate at a spreading center.

These oceanic crustal rocks or ophiolites, are best exposed in the seacliffs both north and south and east of the Point Sal headland. The ophiolite includes gabbro (Figure 37), serpentine derived by alteration of basalt, and peridotite, a rock composed chiefly of the green, iron, magnesium, silicate mineral called olivine. Olivine is sometimes a gem mineral, but sadly, the olivine at Point Sal is too fine grained and too deeply weathered to be of any value to a mineral collector.

Another band of ophiolitic rocks occurs in the Figueroa Mountain area, but is not as complete or as well exposed as the Point Sal sequence.

Point Sal, incidentally, was named by the British explorer, George Vancouver in 1792, in honor of Hermenegildo Sal, then Comandante at San Francisco.

Mussel Rock

This is the last notable headland in the county, two miles north of Point Sal. It is yet another headland formed from a relatively resistant exposure of the Monterey Formation. This headland, like others to the south, is capped by a sheet of dune sand that extends inland more than a mile.

CHAPTER 3

THE ROCKS AND GEOLOGIC HISTORY OF SANTA BARBARA COUNTY

Each of the three great classes of rocks, **sedimentary**, **igneous** and **metamorphic**, is represented in Santa Barbara County. On the mainland, however, sedimentary rocks are by far the most often encountered. However, both igneous and metamorphic rocks are present in a band from near Figueroa Mountain to the area of Gibraltar Reservoir, at Point Sal, in the Cuyama River gorge and at a few other places. In contrast, on the offshore islands, igneous and metamorphic rocks are much more prominent.

Sedimentary rocks are those that form on the surface of the earth, whether on land or on the sea bottom. Most are deposited as muds, sands or gravels, that have been eroded from higher areas and transported to the depositional site typically by streams, but sometimes by ice or wind. (There are no ice-transported materials in any part of the county.) Geologists call these transported sediments **clastic** to distinguish them from **chemical sediments** such as lime, salt and some other materials that can be precipitated directly out of solution, and from **organic sediments** that are composed largely or wholly of organic material. Coal is perhaps the best known organic sedimentary rock, but we have virtually no coal in the county. However, we do have extensive deposits of diatomite, a relatively rare sedimentary rock composed of the remains of microscopic green plants called diatoms. Diotomite will be discussed in greater detail later in this chapter.

Loose sediment such as sand, mud or gravel, becomes sedimentary **rock** only when the grains are cemented or pressed together. Sands and gravels require cementation to become rock; muds, silts and organic materials may become rock when their grains are pressed tightly together.

Igneous rocks are those that have solidified from a molten state and are usually divided into two main categories: **volcanic** where solidification takes place on or near the earth's surface, and **plutonic** where the molten material congeals while deeply buried.

The different cooling environments lead to distinctly different characteristics. Volcanic rocks exposed to the atmosphere lose heat rapidly and cool quickly so that mineral crystals that form them have little opportunity to grow. Plutonic rock, on the other hand, forms well below the earth's surface and cools very slowly, permitting mineral crystals to grow much larger. As a result, volcanic rocks are typically fine-grained, and plutonic rocks coarse-grained enough so that it is easy to see individual mineral grains.

Some volcanic rocks show what is called "pillow structure", in which the rock forms rounded masses about the size and shape of a stack of pillows. We have learned that this structure is developed when lava is erupted under water. It could develop in a lake, but the usual site would be the sea floor (Figure 37).

The third great class of rocks are those called metamorphic. These are rocks that have been altered short of melting by heat, pressure or chemical solutions. Their original nature is sometimes difficult to deduce if the alteration is intense enough. They may be derived from sedimentary, igneous or even metamorphic parents. There are many degrees of metamorphism. These range from only a modest change such as the alteration of a sedimentary shale to a metamorphic slate, through to extreme metamorphism that nearly melts the rock, often producing a product that is difficult to tell from a truly igneous rock like a granite.

The greenish serpentinites that occur in the Figueroa Mountain area are metamorphic rocks that have resulted from the alteration of sea floor volcanic rocks, usually basalts.

It is from many sorts of information teased out of rocks that geologists are able to put together a geologic history of a given area. Not all rocks are useful for tracing the history of life or the nature of past climates or other environmental conditions on the earth's surface, but they may reveal other essential information for reconstructing the whole geologic history. Sedimentary rocks, for example, may tell us about past climates, what sorts of plants and animals were present by recording or revealing such things as the location of rivers, lakes and coastal plains. If we arrange these sedimentary rocks in order of their age, we can provide a framework for a geologic history.

The three **columnar sections** (Figures 38 to 41) show the age-order and the relative thickness of the rock units present in three representative areas of the county. Columnar section represent the sequence of rock layers present in an area, in stratigraphic order, that is, with the oldest the bottom and the youngest at the top. These diagrams customarily show the relative thickness of each of the units and indicate by conventional symbols, the rock type.

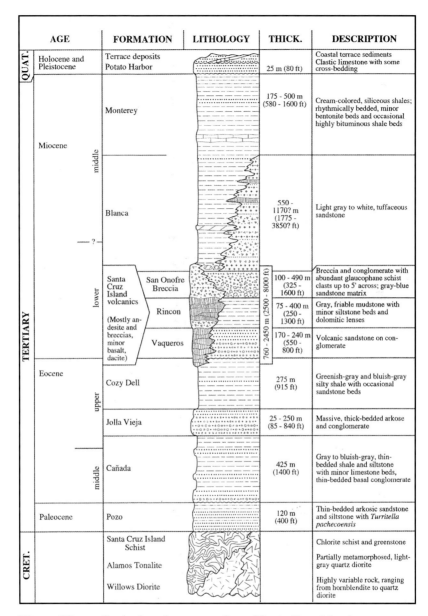

Figure 38. Geologic columnar section of Santa Cruz Island.

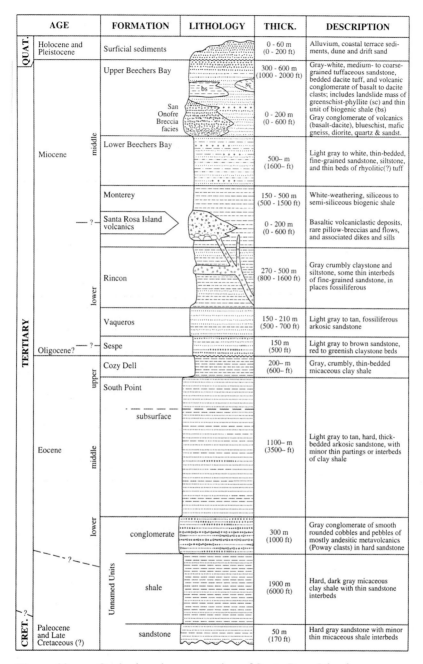

Figure 39. Geologic columnar section of Santa Rosa Island.

AGE		FORMATION	LITHOLOGY	THICK.	DESCRIPTION
Recent		Alluvium		0-100'	Silts and gravels
Pleistocene	upper	Terraces		0-100'	Gravels
Pliocene	lower	Sisquoc		3200'+	Diatomaceous siltstone Clay shale or diatomaceous mudstone
	?				
Miocene	upper	Monterey		1000' - 3000'	Thin-bedded clay shale or laminated diatomite Porcelaneous and cherty siliceous shales Organic shales and thin limestones
	middle				
	lower	Tranquillon		0-1200'	Rhyolite and basalt lava, agglomerate, tuff, bentonite
		Rincon		0-1700'	Claystone
		Vaqueros		0-900'	Sandstone and conglomerate
Oligocene		Sespe / Alegria		0-2000'	Pink to buff sandstone and red and green siltstone Gray to buff marine sandstone
		Gaviota		1600'+	Fossiliferous buff sandstone and siltstone
Eocene	upper	Sacate		1000' - 1500'	Buff sandstone and clay shale
		Cozy Dell		700' - 2000'	Brown clay shale
		Matilija		0'-2000'	Buff arkosic sandstone
	middle	Anita		0-1000'	Dark gray clay shale
		Sierra Blanca		0-50'	Algal limestone lens
Cretaceous	Upper	Jalama		2200'+	Buff fine-grained sandstone Gray siltstone Buff sandstones and grey clay shales
	middle ? and Lower	Espada		4000'+ to 6800'+	Dark greenish brown carbonaceous shales and thin sandstones Basal pebbly sandstone
	?				
Jurassic	Upper	Honda		1500'	Dark greenish brown nodular claystone
		Franciscan		?	Hard green sandstone and black shale Serpentine intrusions

Figure 40. Geologic columnar section of the western Santa Ynez Mountains. (after T. W. Dibblee, Jr.)

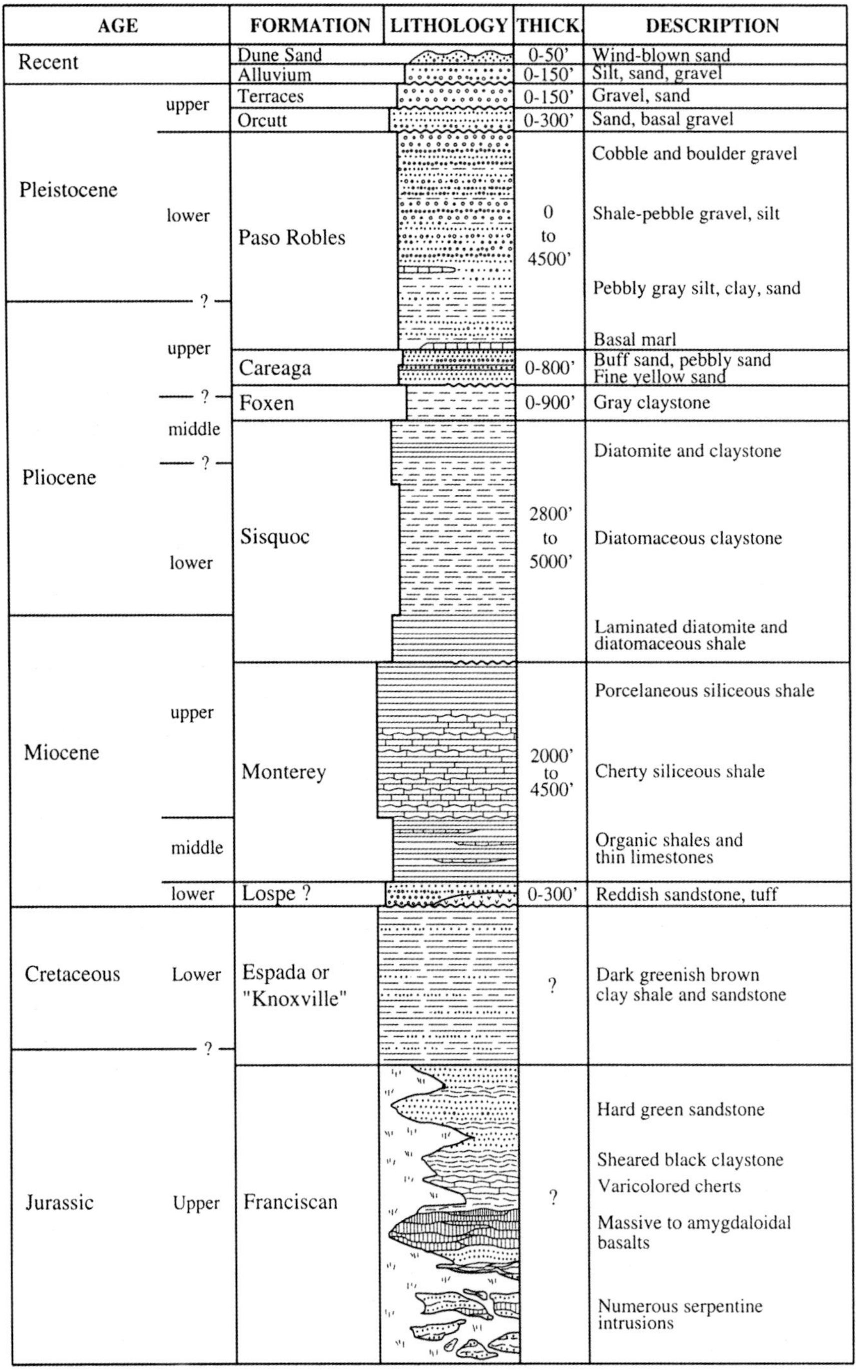

AGE		FORMATION	LITHOLOGY	THICK	DESCRIPTION
Recent		Dune Sand		0-50'	Wind-blown sand
		Alluvium		0-150'	Silt, sand, gravel
Pleistocene	upper	Terraces		0-150'	Gravel, sand
		Orcutt		0-300'	Sand, basal gravel
	lower	Paso Robles		0 to 4500'	Cobble and boulder gravel
					Shale-pebble gravel, silt
	— ? —				Pebbly gray silt, clay, sand
Pliocene	upper	Careaga		0-800'	Basal marl
					Buff sand, pebbly sand
					Fine yellow sand
	— ? —	Foxen		0-900'	Gray claystone
	middle				Diatomite and claystone
	— ? —	Sisquoc		2800' to 5000'	Diatomaceous claystone
	lower				Laminated diatomite and diatomaceous shale
Miocene	upper	Monterey		2000' to 4500'	Porcelaneous siliceous shale
					Cherty siliceous shale
	middle				Organic shales and thin limestones
	lower	Lospe ?		0-300'	Reddish sandstone, tuff
Cretaceous	Lower	Espada or "Knoxville"		?	Dark greenish brown clay shale and sandstone
	— ? —				
Jurassic	Upper	Franciscan		?	Hard green sandstone
					Sheared black claystone
					Varicolored cherts
					Massive to amygdaloidal basalts
					Numerous serpentine intrusions

Figure 41. Geologic columnar section of the southern Santa Maria Basin.

Columnar stratigraphic sections are generalizations and in a given area not all rock units may be present at every location. Variations in thickness or lithology can be expected. Some rock units, such as the Miocene Monterey Formation were deposited over such a large area that this unit appears in each of the three areas. On the other hand, some rock units are represented in only one or two columns. The reasons for these differences will become clearer as the nature of the various rock units is discussed in the ensuing section.

Igneous and metamorphic rocks do not tell us much about climate or the history of life, but they may reveal a great deal about where past mountain ranges were located or about the interactions of the great crustal plates. During the past century, for example, we have found convincing evidence that shows that continents moved, albeit rather slowly, with respect to one another. Some now in tropical areas were in polar regions in the past. As noted earlier, we call this movement **continental drift**.

Much of the evidence for continental drift (or for smaller scale movements of parts of continents), has come from volcanic rocks that contain iron-bearing minerals. When the volcanic material is still a hot fluid, iron-bearing minerals line themselves up parallel to the earth's magnetic field at that time and place. These "fossil compasses" are locked into position when the lava congeals to a rock. If these rocks are shifted by continental drift or other process to some new position, the fossil compasses may no longer point to the earth's magnetic pole. If we can date these volcanic rocks, we can put together a history of which way the area moved and how rapidly it took place. This technique, for example, has been used on volcanic rocks from the islands off the south coast. Surprisingly, it shows that the island chain has been rotated 60-80° in a clockwise direction since the lava flows were erupted in Miocene time, something like 15 million years ago. The rotation appears to be due primarily to displacements on the San Andreas fault which have jostled and shifted southern California's crustal blocks and played a major role in creating the distinctive orientation of the Transverse Ranges.

Because most of the rocks exposed in Santa Barbara County are sedimentary, they necessarily provide most of the evidence we need to reconstruct the geologic history of our area. A number of these sedimentary units or **formations** have already been mentioned as we considered the landscape features of the county. In this chapter we will give these rock units a closer look.

A formation is a rock unit whose distinctive characteristics allow it to be readily recognized and its distribution plotted on a map. It must be thick enough to be shown on a map and extensive enough to be followed perhaps a mile or so across the country. A 1-inch thick layer of limestone exposed in a canyon or two would not usually qualify. The distinctive characteristics of a formation might be its composition or color, a sandstone for example, or perhaps something like its resistance to erosion. In short, anything that makes it readily distinguishable from the other rocks with which it is associated. Although it must be thick enough to be plotted on a map, it may range from a few feet to thousands of feet in thickness. Because formations are usually deposits that have formed in a specific environmental setting such as a beach, delta or offshore basin, their extent has limits and if one follows them across country, they will eventually get thinner and thinner, or "pinch out" as geologists say. In some cases they may grade into another different rock type or formation, the sort of thing you'd expect when looking at an ancient beach deposit. In one direction, this ancient beach would be expected to grade into a shallow water, near-shore marine deposit. In the opposite direction it would be replaced by some sort of nearshore land deposit.

We often refer to the **geologic age** of a formation (Figures 38, 39 and 40), but age is not something we can determine by just looking at the appearance of the formation; rather we must use something like its contained fossils or perhaps its relationship to other formations whose age has already been established. In short, the age of a rock is not a visible characteristic.

You have already become aware that formations have names, often geographic names. We have mentioned the Monterey Formation on a number of occasions. Its name was assigned by the first geologist to write and publish a formal description and to select a **type locality**, a place where the rock unit is especially well displayed. In the case of the Monterey Formation, the type locality was in the hills south of the City of Monterey.

Where a formation like the Monterey is found to occur in Santa Barbara County some distance from the type locality, geologists have either traced it continuously across the intervening area, or established that it is the same unit on the basis of similarity of rock type, color, relation to other rocks or by fossil content. This second method is the one that must be used in assigning the name to rocks on the offshore islands. Of course, now and then these long-distance correlations prove to be wrong as new evidence is found that proves to be more conclusive.

Before we look at the rock units and geologic history in detail, it is well to point out that Santa Barbara County's geologic record is not very long as geologic time goes. The oldest rock we have found anywhere on earth is a little less than 4,000 million (4 billion) years old. The oldest rocks in Santa Barbara County are only about 150 million years old, pretty elderly to be sure if compared to human history, but as far as the earth's history goes, about 96 or 97 percent of it had elapsed before our oldest rocks were even formed.

Putting it in other terms, the county's oldest rocks were formed from sedimentary and volcanic material deposited during the heyday of the great dinosaurs. Perhaps unfortunately, no remains of these creatures have been found within the county and it is very unlikely that any ever will turn up for the very good reason that our rocks of appropriate age are either marine deposits or rocks that have been so thoroughly metamorphosed that any traces of fossils have been destroyed. It is possible, certainly, that a fossil of one of the large marine reptiles might be discovered, but probably not a land-living dinosaur.

JURASSIC AND OLDER ROCKS

Santa Cruz Island Schist

The oldest rock in Santa Barbara County is the Santa Cruz Island schist, a metamorphic rock with a crude, sheet-like structure, often rich in platy minerals like the micas. This rock is greenish-gray and rich in the dull green, micaceous mineral chlorite. Where it has been exposed to a long period of weathering, it tends to form a brick-red soil the result of the alteration of various iron-bearing minerals. The schist can be seen in a band about 10 miles long on the crest of the ridge south of the island's Central Valley.

This rock itself has not yet been dated, but we do know that a younger igneous rock, the Alamos tonalite, has intruded it, and we know that the Alamos tonalite is about 145 million years old, demonstrating that the intruded schist is still older, though how much older is uncertain.

Franciscan Formation

For many years this rock unit, very widespread in the California Coast Ranges from Santa Barbara County north into Oregon, had been a real enigma to geologists. Our understanding is still imperfect, but we do now have a much clearer idea of how and where it formed. It includes a number of bodies of different kinds of rock such as dark-colored sandstone (graywacke), layered, usually reddish, cherts of deep sea origin (Figure 42), serpentinites,

Figure 42. Blue Canyon Pass in the Santa Ynez Mountains. Thin-bedded and deformed radiolarian chert of the Jurassic-Cretaceous Franciscan Formation.

intrusive gabbro, blue glaucophane schist, pillow basalt, dikes and sills. The potpourri of rock types that make up the Franciscan Formation do not have any systematic or predictable relationship to one another. Instead they occur, often, as discrete blocks separated from one another by faults. This jumbled mixture of rock types is called a **melange**, a fancy name for a medley of rock types. Fossils are very rare so that attempts to work out a history of the rock or fit it into the history of the California Coast Ranges was frustrating in the extreme.

As the theory of Plate Tectonics took shape in the 1960's and 1970's, it gradually became evident that the Franciscan Formation consisted of an assemblage of scrapings from the deep sea floor that were mixed and plastered against the continent and in some cases dragged into and later disgorged from a subduction zone as the oceanic plate moved beneath the North American continental plate. The size of the individual rock masses within the Franciscan varies from a few feet to large masses or blocks several miles long.

Judging from the few fossils that are found in the Franciscan, by the younger, undisturbed rocks that were deposited on top of the Franciscan and by other field relationships, we know that his formation was formed from late Jurassic time until perhaps the early Eocene (165 to 45 million years ago).

Figure 43. Figueroa Mountain, south face in November 1995. The late Helmut Ehrenspeck at the site of the fossil "black smoker" in the Jurassic-Cretaceous Franciscan Formation.

The unusual suite of rocks that comprises the Franciscan Formation is displayed in a prominent band in the San Rafael Mountains southeastward from Zaca Creek (Figure 2). The Little Pine fault marks the southwest side of this slice of rock and the Camuesa fault the northeast side. This belt of rock pinches out near Gibraltar Reservoir on the Santa Ynez River. To the north, the Franciscan assemblage is also exposed in the gorge of the Cuyama River.

Other small examples of the Franciscan Formation can be seen on the north flank of the Santa Ynez Mountains from Blue Canyon east to Juncal Dam. Another is present on the north slope of Figueroa Mountain. Interestingly, the Franciscan rock at Figueroa Mountain includes some small bodies of chromite, the principal ore of chromium, which from time to time have supported short-lived, small-scale mining operations.

One of the more remarkable features of the Franciscan belt of rocks south of Figueroa Mountain is a fossil "**black smoker**", now regrettably vandalized (Figure 43). Black Smokers are high-temperature submarine steam vents that occur along active spreading centers on the sea floor, first discovered in the 1980's off the Pacific coast of Central America. These submarine, chimney-like structures emit clouds of black metallic sulfides and become encrusted and eventually clogged by these sulfide minerals. These bodies, at various times and places, have led to the development of important ore bodies such as the copper ores on the island of Cyprus. They are also remarkable because of the large tube worms, mollusks and other organisms that cluster around them and are sustained by geothermal rather than solar energy; they live in what Dr. Rachel Haymon of UCSB has likened to a toxic waste dump.

The Figueroa Mountain example includes a small copper deposit, pillow lavas, and fossils of organisms that look very much like the life that surrounds modern black smokers – despite the fact that this Figueroa Mountain example is over 100 million years old.

Two additional examples of Franciscan rocks occur in the county, one on the south side of the Honda fault in the westernmost Santa Ynez Range southwest of Lompoc in the Honda Creek drainage where a clay shale is associated with serpentinite.

The other exposure is at Point Sal and was described in the section on that headland.

Although exposure of this interesting group of rocks is confined only to certain areas in the county, it is very likely that much of the county, except perhaps the offshore islands, is underlain at depth by such rocks, or as

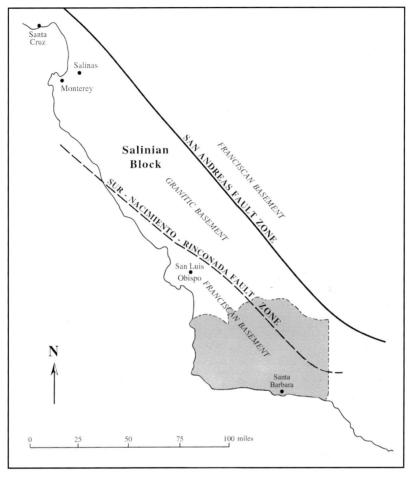

Figure 44. Map of the central coast of California showing the distribution of basement rock types in the southern Coast Ranges.

geologists say, the Franciscan forms the **basement** except for the areas northeast of the Naciminento-Rinconada fault zone which probably has a Sierran-like granitic basement now deeply buried beneath sedimentary rocks. This is likely because not far to the north, in San Luis Obispo County, such granitic rocks can be seen poking through the sedimentary cover. This slice of rock with a granitic basement lies between the Naciminento-Rinconada and San Andreas fault zones and is called the Salinian Block (Figure 2 and 44).

EARLY CRETACEOUS ROCKS

Honda and Espada Formations

These two rock units have quite a similar appearance. Both are brown shales with minor sandstone beds and both have a faint greenish cast. They are about the same age, though in the Honda Creek area near Lompoc, the Honda Formation lies below the Espada and is separated from it by an unconformity, a break in the record during which some erosion of the Honda Formation took place. This requires at least a brief period of uplift (or drop in sea level) because both rock units are marine deposits. Neither rock unit is very fossiliferous, so we lack good age control, but what evidence we do have suggests that the Honda Formation was deposited about the same time as part of the Franciscan. In some places in the Santa Ynez Range, the Espada Formation appears to grade into the late Cretaceous Jalama Formation.

The only place one can see the Honda Formation is on the north wall of Cañada Honda south of Lompoc. The Espada Formation, on the other hand, is much more extensive. It is present on the south and east sides of Tranquillon Mountain, along El Jaro Creek, and on either side of U.S. Highway 101 from the Santa Ynez River southward for about a mile. In the San Rafael Mountains, there are many exposures south of Manzana Canyon southeastward to beyond Gibraltar Reservoir (Figure 45). These rocks and those parts of the Franciscan Formation of equivalent age originally may have been deposited in widely separated parts of the marine environment, but have since been brought into close proximity by sea floor spreading, subduction and faulting. You will remember that the Franciscan Formation includes a lot of deep water deposits and was deposited some distance offshore. The Espada, on the other hand, so far as we can tell by its few fossils, was deposited much closer to the shore. As we have unraveled the very complicated history of the California Coast Ranges, we have found that some rock units now in contact with one another were originally deposited in basins hundreds or even thousands of miles apart, but were later moved by plate motions and faulting into their present position.

LATE CRETACEOUS ROCKS

By 80 or 90 million years ago, in the latter part of the Cretaceous Period, the accumulation of Franciscan Formation deposits pretty well ended so far as we can tell from Santa Barbara County rocks and we find no more rocks

Figure 45. Little Pine Mountain, south face. Marine Cretaceous Espada Formation (smooth slopes) capped by the harder marine Tertiary Temblor Formation. The high cliffs in the distance are marine Miocene Monterey shale.

obviously associated with subduction until the Miocene when volcanic activity became very prominent throughout California.

However, the period of marine conditions that had begun earlier when the Honda and Espada formations were deposited, continued more or less without a break until Eocene time, about 40 million years ago. During this 40-50 million year period, there was little or no land in what is now Santa Barbara County. The Mesozoic Era and the Cretaceous period ended with a bang when a large asteroid or meteorite crashed into the earth near the Yucatan Peninsula of Mexico. Large reptiles, like the dinosaurs, as well as many less spectacular animals became extinct and the Cenozoic Era and Paleocene Epoch were ushered in. Mammals soon became the dominant land animals.

All this geologic drama left scarcely a trace in the Santa Barbara County rock record. Marine deposition apparently continued without interruption. So little change occurred that sediments deposited in the final days of the Cretaceous look almost exactly like those laid down as the Cenozoic and its Paleocene Epoch began about 60 million years ago. So similar are the rocks deposited across this time boundary, that in the Sierra Madre in the north county, the same rocks have been mapped by some geologists as late as Cretaceous and by others as Paleocene. This is understandable not only because deposition occurred without a break across this time boundary, but also because the rather monotonous sequence of sandstone and shale continued without change and because diagnostic fossils are few and far between.

When an event like an asteroid impact occurs, it usually leaves a signature in the huge amount of dust stirred up into the atmosphere which will promptly be distributed around the world. The signature is an abrupt increase or "spike" in the amount of the trace element iridium. The iridium spike is confined to such a thin layer, that although it is probably present somewhere in the rocks of Santa Barbara County, but because of the enormous pile of sediment that accumulated during that 40-50 million year period means that finding the spike is apt to be more difficult than finding the needle in the proverbial haystack.

Although widespread and very thick, this late Cretaceous to Eocene sequence of rocks is not very exciting. It does not lead to dramatic landscape features, it looks about the same from one place to another and offers few opportunities for the fossil collector.

It occurs in the San Rafael Mountains, on either side of the Manzana Creek, at McKinley, Cachuma and Big Pine mountains, over much

Figure 46. Sierra Madre, crest of range near Bates Canyon. Marine Paleocene and Cretaceous rocks, composed chiefly of sandstones.

of the upper Sisquoc River drainage north of the Hurricane Deck, and in the Sierra Madre mountain range (Figure 46).

Slightly different, sandier rocks of late Cretaceous age are found in upper Salsipuedes Creek near Lompoc and along Jalama Creek southward to upper Santa Anita Canyon on the south face of the Santa Ynez Range. At Jalachichi Summit on the Jalama Road there are some good exposures of these sandstones. Other patches are found on the north slope of the Santa Ynez Range near Forbush Flat and Blue Canyon.

TERTIARY (PALEOCENE & EOCENE) ROCKS

We have seen that the late Cretaceous and Paleocene rocks in the north county form a continuous sequence lacking any obvious breaks, but on Santa Cruz Island there is one rock unit, the Pozo Formation found in the southwestern part of the island, that contains abundant and definitive Paleocene fossils.

Sometime during the period of 40-50 million years ago, the sea withdrew from what is now Santa Barbara County and land erosion took over until middle Eocene time about 40 million years ago. At this time, the sea advanced over the area and marine deposition was resumed. This erosional interval could have been as long as 10 million years or perhaps as little as one million years. We know only that the youngest rocks that escaped being eroded are about 50 million years old and that the oldest deposited after the erosional interval are about 40 million years old. This is a common problem whenever geologists attempt to construct a geologic history from rock sequences where erosion has destroyed much of the evidence. It also means that history is provisional and may be revised as better dating methods or new rock evidence is discovered.

Sierra Blanca Formation

The oldest of the Eocene rock units in the San Rafael Mountains and adjacent areas is the Sierra Blanca limestone. It is generally not very thick nor very extensive, but it is a hard, ridge-former, nearly white and very prominent where it occurs. Further, it is the most widely-distributed limestone in the county, a rock type relatively rare in our region (Figure 47).

It is what is called a clastic limestone, a rock composed of broken fragments of calcareous organisms including algae, corals, worms, foraminiferans and some brachiopods mixed with sand and well-rounded pebbles. It was deposited in a shallow water, near-shore, warm environment, probably shoreward from a fringing coral reef—a setting considerably more tropical than prevails today.

The Sierra Blanca weathers to a white color and as a result stands out in contrast to the surrounding rocks that usually are tan or brown. It is easily seen on the north wall of Blue Canyon near Forbush Flat, as well as at many points from East Camino Cielo Road. The thickest, best example of this distinctive rock is at its type locality about seven miles north of the junction of Indian and Mono creeks where it is several hundred feet thick.

Anita Formation

In the western part of the Santa Ynez Range, the Anita Formation is the same age as the Sierra Blanca because enclosed within the Anita shales are two small lenses or beds of typical Sierra Blanca limestone. One of these is in Jalama Canyon and the other in lower Nojoqui Canyon.

Good exposures of the Anita Formation can be seen on the Hollister Ranch west of Gaviota. The Anita is mostly a red and green shale containing very

Figure 47. Santa Ynez Mountains, just north of Blue Canyon Pass. Marine algae limestone of the Eocene Sierra Blanca Formation.

abundant, fairly large foraminiferans. Most forams can be seen only with a hand lens or microscope, but those of the Anita can be seen with the naked eye. Oil geologists have long been impressed with the abundant foraminiferans in this rock and have informally called it "The Poppin Shale".

Juncal Formation

In Agua Caliente Canyon, a little north of the Juncal Forest Service Campground, is an exposure of the lower Juncal Formation. The Juncal rests on the Sierra Blanca and in this area contains fossil oysters. Oysters are good indicators of shallow marine, estuarine conditions often associated with river mouths.

While both the reef deposits of the Sierra Blanca and the oyster-bearing shales of the younger Juncal Formation indicate nearshore conditions, they point to very different environments. Both the Anita and Juncal formations show that erosional and depositional processes can change fairly abruptly.

Although clear-water reef, conditions prevailed only during deposition of the Sierra Blanca limestone, younger Eocene rocks show a persistent alternating pattern between sandstone and shale deposition. It is a truism that coarser sediments like sands and gravels will be deposited near shore where waves and currents will keep the finer-grained muds and silts suspended in the water. Farther offshore, beyond the reach of breaking waves and coastal currents, muds and silts will have an opportunity to settle out of the water column.

Matilija, Cozy Dell and Coldwater Formations

The alternating pattern just referred to is very well displayed in the Santa Ynez Mountains where resistant sandstones form ridges and prominent bare rock exposures and the softer shales and mudstones form swales and valleys (Figures 40 and 48). Even within the generally shaly units like the Cozy Dell Formation, there are some sandstone beds that suggest changes in the depositional environment, probably a shift toward shallower waters. One such example is the ripple-marked sandstone bed within the Cozy Dell indicating that the water was shallow enough so that the bottom sediment could be agitated by waves (Figure 8).

The Matilija Formation forms the range crest at Montecito Peak, La Cumbre Peak and farther west at Broadcast and Santa Ynez peaks, but at

Figure 48. Santa Ynez Mountains; view west from the Tunnel Trail. Marine Eocene Coldwater sandstone forms the ridge on the left. The softer, marine Eocene Cozy Dell Formation underlies the swale, and the more resistant marine Eocene Matilija Formation forms the ridge on the right.

San Marcos Pass it is the younger Coldwater sandstone that forms the summit area.

Westward from San Marcos Pass, the sandstones of the Coldwater are gradually replaced by the sandstones and shales of the Sacate Formation, an appreciably less resistant rock. The Sacate forms the ridge crest at Refugio Pass.

The Matilija Formation, however, persists as a recognizable unit as far as Point Conception, but it gets progressively thinner and less prominent.

Not only do these two resistant sandstones account for most of the bare rock exposures in the central and eastern Santa Ynez Range, but they are also the source of the huge sandstone boulders that are so prominent in the creeks and lower foothills. It boggles the mind to imagine the sort of catastrophic events that delivered these enormous rocks down the canyons. Almost certainly they were moved by large-volume, high-intensity mud flows because water alone would not be equal to the task.

Boulders from both formations look so much alike, it is virtually impossible to tell which formation produced a given boulder.

During the balance of the Eocene Epoch that ended about 39 million years ago, the sea remained over the southern and eastern parts of what is now Santa Barbara County leaving behind thick piles of sediments that are so well displayed in the Santa Ynez Mountains. However, if we trace these Eocene deposits northward, they rapidly become thinner and are not even present in the Santa Maria area; they may never have been deposited there. The upshot of this is a very long gap in the rock record of the north county, from Jurassic Franciscan deposition to the Miocene, land-deposited Lospe Formation.

TERTIARY (OLIGOCENE) ROCKS

Sespe, Gaviota and Alegria Formations

The Oligocene Epoch began about 39 million years ago and came to a close about 23½ million years ago. It is an interesting time in the geologic history of Santa Barbara County and much of coastal southern California because what had long been a marine environment became a dry land one. We are not completely certain about the cause or causes of this change, but we do know that our region underwent a period of tectonic uplift and deformation—the Ynezan orogeny. Although this **orogeny** was probably the chief cause of our local change from sea bottom to land, there are many places in the world where we have evidence that sea level actually dropped during Oligocene time, likely caused by an expansion of continental glaciers on Antarctica. It therefore seems likely that the change we observe in the Santa Barbara area is the result of both tectonic events and sea level change.

What sort of evidence tells us that the sea was beginning to withdraw from the land?

In the mountains north of Santa Barbara, along both Gibraltar Road and San Marcos Pass are several rock exposures that show clearly a change in environment was on the way. As is usual with such geologic events, the transition from marine to land conditions was not a smooth, unbroken change, but rather a fluctuating one, characterized by some brief reversals. This is easy to see in roadcuts on Gibraltar Road near the Mount Calvary Retreat and along Old San Marcos Pass Road below its junction with State Highway 154. In examining this evidence, remember that the rocks get older as one goes uphill.

Figure 49. Gibraltar Road near Santa Barbara. Nearly vertical pebble conglomerate in the lower part of the nonmarine Sespe Formation, mostly of Oligocene age.

Just north of the Retreat is a low divide between Sycamore Canyon on the east and Rattlesnake Canyon on the west. Along the road a short distance uphill from this divide is the Coldwater Formation. In it are several thin reddish layers showing that brief retreats of the sea were followed by brief advances in which the yellow-brown marine beds covered the land-laid red beds. Red sedimentary rocks are most often indicative of land deposition because the iron-bearing minerals in them have had more opportunity to oxidize and convert these oxides from yellow minerals like limonite to red ones like hematite.

Downhill from the divide, the younger lower Sespe Formation is a pebbly, dark red stream deposit showing that full withdrawal of the sea occurred (Figure 49) prior to its deposition.

Further, if one looks carefully at the top of the Coldwater Formation in that area, additional evidence can be seen of a nearshore environment, namely fossil oysters. Oysters inhabit estuaries along coastlines. Similar evidence can be seen on Old San Marcos Pass Road near the tight switchbacks.

The Sespe Formation got its name from an extensive exposure in upper Sespe Creek in central Ventura County where it was first described. Other exposures occur as far east as Los Angeles County and as far west as Gaviota Pass. For many years the Sespe was thought to be nearly unfossiliferous, particularly in Santa Barbara County, and its age was determined by fossils in the older rocks below and in younger ones above. However, in recent years, particularly in the vicinity of Simi Valley in eastern Ventura County, about 120 species of vertebrates have been discovered, a good many of which are represented only by teeth, not bones.

In Santa Barbara County, in the lower part of the Alegria Formation near Gaviota Pass, teeth of an oreodont have been found. The Alegria is the marine equivalent of the Sespe, so it is clear that the fossil teeth of this land mammal were washed into the sea from the nearby land. Oreodonts were pig-like animals that roamed open country in large herds. They are very well known from rocks in the Mississippi and Missouri river valleys and range in age from late Eocene to Pliocene.

Among the 120 or so vertebrate animal remains found in Sespe rocks in Santa Barbara and Ventura counties are crocodiles, horses, camels, sabre-toothed cats, marsupials like the opossum, lizards, tortoises, mouse-like rodents and many others. Fossils of invertebrates like snails and freshwater clams have also been recovered from Sespe rocks.

Measurements of the alignment of magnetic mineral grains in the Sespe tells us that the North American land mass was then much farther south than it is today. The Santa Barbara area was then located about 17-18° North, and almost certainly had a more tropical climate. The dark red color of the lower Sespe is further indication that deposition occurred in a tropical setting because today we find brick-red sediments called laterites in tropical settings where markedly alternating wet and dry seasons occur. Sediments deposited in the sea, by contrast, tend to be dark-colored greenish to gray or even black because there is insufficient oxygen in seawater to oxidize the iron minerals to yellows and reds. Hence when freshly exposed, as in a new deep road cut, these rocks are often bluish or greenish gray. Once these marine rocks are exposed to the atmosphere, they slowly oxidize and develop a tan or yellow-brown color as the iron mineral limonite forms from the darker, unoxidized iron minerals. Limonite, incidentally, has about the same composition as ordinary iron rust.

In addition, if we look closely at the layers or strata in the Sespe, we find that they are full of small gravel-filled channels virtually identical to those seen in modern streams and flood plains.

Detailed investigations of the Sespe Formation have demonstrated that it ranges in age from about middle Eocene through the Oligocene to early Miocene (approximately 40-20 million years ago). Further, it generally can be divided into two parts. The older, lower one is characterized by sediment-choked braided stream and delta deposits. This was followed by what appears to be a long erosional break, probably embracing much of the Oligocene Epoch. The upper, younger part of the Sespe was deposited in both braided streams and in sluggish, meandering rivers carrying more mud and silt than gravel.

The distinctive red Sespe Formation is exposed on the south face of the Santa Ynez Mountains from the Ventura County line westward to Gaviota Pass. However, near Glen Annie Canyon, the upper part of the Sespe begins to grade into and be replaced by the marine Gaviota Formation which is quite fossiliferous and very prominent at Gaviota Pass. The Gaviota is, in turn, replaced westward by the marine Alegria Formation. These relationships show us that the Sespe is approximately the same age as the Gaviota and Alegria formations and that it passes laterally into a marine environment of deposition. If one were to stand on the boundary between the Gaviota and Sespe formations, one would be on the exact location of an ancient beach.

The well-rounded pebbles found near the bottom of the Sespe formation are very interesting and provide clues to the nature of the local geography when they were deposited. Many, perhaps most, are hard, very well-rounded quartzites and metamorphosed volcanic rocks, neither of which can be traced to any sources either in the Sierra Nevada or in the Coast Ranges; they had to come from somewhere else. Mostly likely there are some of the desert mountains in the Mojave Desert which have beds of quartzite and metavolcanic rock. There is also a slight possibility that some of these hard pebbles in the Sespe have been eroded out of some older pebbly conglomerate and are, in effect, just recycled pebbles whose ancestry is quite obscure.

Among the Sespe pebbles are some granitic rocks that may have come from the Sierra Nevada, the Mojave Desert Ranges or the eastern Transverse Ranges near San Bernardino. They might have come from even closer sources now buried by younger rocks. The truth is that we really don't know where they originated. It is very unlikely they are recycled because granite, unlike quartzite, disintegrates fairly rapidly during transport, certainly much faster than either quartzites or metavolcanic rocks.

An interesting exposure of the Sespe Formation containing granitic pebbles and small boulders is located on East Camino Cielo near the Painted Cave Road. Both granitic and pebbles from the Franciscan Formation are present here. The granitic pebbles look very fresh from a few yards away, but turn out to be so thoroughly weathered that one can stick a pen knife into them quite easily – but don't try this with the Franciscan cherts, the quartzites or the metavolcanic pebbles.

Other exposures of the Sespe Formation can be seen south of Solvang and northwest of Nojoqui Summit on U.S. Highway 101 where the pebbles are almost entirely from Franciscan sources. The Sespe can also be seen south of the Big Pine fault in the San Rafael Mountains near the Ventura County line, and on the south face of the Santa Ynez Mountains near Carpinteria where the chaparral has been cleared and avocados planted on the red soil.

Three small exposures of reddish Sespe are present in the Santa Barbara city area, one near the entrance to Hope Ranch on the south side of Modoc Road, another along Loma Alta Drive on the steep hillside below the TV station, and a third patch on a low hill on San Pascual Street between Ortega and Cota.

Other sedimentary rocks resembling the Sespe, and of about the same age, occur at several places in the county. On the offshore islands, a Sespe-like rock is present on the southern part of Santa Rosa Island and was encountered in an exploratory oil well drilled at the eastern end of Santa Cruz Island.

Chapter 3

Figure 50. Shells of the large, robust scallop (*Lyropecten magnolia*) of early Miocene age are commonly found in the Vaqueros Formation.

In the Cuyama Valley area, both in the Sierra Madre and across the river in the Caliente Range of San Luis Obispo County are numerous exposures of the Simmler Formation, a non-marine rock composed of sandstones and shales of reddish, greenish and gray color, similar in age and appearance to the Sespe Formation.

TERTIARY (MIOCENE) ROCKS

This epoch lasted from about 23.5 to 5 million years ago, and the rocks formed during that time are among the most widespread, interesting, and economically important of any to be seen in Santa Barbara County.

The Vaqueros Formation

Near the end of the Oligocene Epoch, the sea began to spread inland over the broad plains and deltas where the Sespe Formation had been deposited. These marine beds are usually coarse, fossiliferous sands, but in a few places they are quite pebbly. The first of these deposits were the beds of the Alegria and Gaviota formations found in the western Santa Ynez Range and were

Figure 51. Gaviota Pass; view west. The very sharp vegetation contact between the grass-covered marine Miocene Rincon Formation and the brush-covered marine Vaqueros and Algeria formations. Rocks in the latter two formation are of early Miocene or late Oligocene age. The high ridge is composed of marine Gaviota Formation of late Eocene age.

formed during late Oligocene time. Later and closer to the end of the epoch, the Vaqueros Formation was deposited. Indeed, a number of geologists claim that the Vaqueros Formation is not late Oligocene, but early Miocene instead. Irrespective of its exact age, it ushered in a period in which all but the northeastern part of the county and a small area near Point Sal, became a marine environment.

One of the characteristic fossils found in the Vaqueros Formation is a large, robust scallop called *Lyropecten magnolia* (Figure 50). Some specimens are nearly six inches across.

The Vaqueros Formation is fairly prominent on the lower south slope of the Santa Ynez Mountains from Summerland westward to Point Arguello. Its prominence comes not from its thickness (it is usually only a couple of hundred feet thick and is much thinner near Point Conception) but rather because of the contrast with the overlying Rincon Formation. The Vaqueros is a hard, ridge-former, but the Rincon is a weak mudstone forming much more subdued topography. The soils that develop on these two rocks are strikingly different. Vaqueros soils are sandy and well-drained; Rincon soils are clayey and poorly-drained. The sandy soils of the Vaqueros support a dense growth of woody chaparral, whereas the Rincon is characteristically grass-covered. The contrast is so sharp on the south face of the Santa Ynez Range that the geological contact can be located within inches by the abrupt change in vegetation. This is spectacularly evident along U.S. Highway 101 between Tajiguas and Gaviota where the change in slope and vegetation is a sharp line. A better example of geologic control of vegetation would be difficult to find anywhere (Figure 51).

Rincon Formation

The Miocene rock record shows that the sea gradually deepened as it spread over much of what is now Santa Barbara County. The nearshore materials of the Vaqueros Formation were deposited as the Oligocene passed into the Miocene. By the time the succeeding Rincon mudstone was laid down, the water was deeper and quieter. The Rincon Formation is distinctive not only by virtue of its soils, but also because it is responsible for more building damage than any other rock unit in the county. Unfortunately, many view properties in the easily accessible lower foothills on the seaward side of the Santa Ynez Range are on this rock unit. Because its soils expand when wet and shrink and crack when dry, it is very destructive to anything built on it unless extensive measures are taken to isolate and protect the structures. Driveways, foundations, pipelines, swimming pools and so on, are apt to be damaged because Rincon soils are in nearly continuous slow motion as a result of wetting and drying. Garden watering, leaky plumbing and septic tanks all exacerbate the problem. Further, Rincon soils usually support shallow-rooted grasses rather than deep-rooted trees, so slopes are very prone to slumping and landsliding following periods of wet weather.

As if these problems weren't enough, the Rincon Formation has been found to be the most prolific source of the radioactive gas, radon, in both Ventura and Santa Barbara counties. In fact, these two counties have the greatest risk of radon exposure in the entire state. Prolonged exposure to high levels of radon have been linked to an increased risk of respiratory cancer. People whose homes are located on the Rincon Formation should have their houses tested for the presence of this radioactive gas. Fortunately, in the generally benign climate of these two counties, most homes are fairly well ventilated which greatly reduces the risk.

Monterey and Sisquoc Formations

After the Rincon mudstone was deposited, a series of very deep, closed basins developed over most of what is now coastal California from San Diego County on the south northward to the San Francisco Bay area. Very poor water circulation prevailed in these basins leading to widespread stagnant or oxygen-deficient conditions. This led to conditions that were toxic to most organisms, especially those that live on the sea floor. As a consequence, sediments settling into these deep basins were not churned up by burrowing animals as they had been when the Rincon mudstone was deposited.

Fossils of bottom-living organisms are therefore extremely rare in Monterey Formation rocks. The fossils that are present represent those animals that lived in the well-ventilated waters well above the bottom and at the sea surface. Foramaniferans, fish skeletons and scales, shark teeth and occasional whale bones are present in this fossil assemblage.

The Rincon mudstone layering is indistinct because sediments were churned by burrowing animals, but in the Monterey, layers are beautifully preserved and often so thin that they represent what was deposited in a single year or season. We call such rocks "laminated" (Figure 22).

For these reasons, the Monterey Formation is a most unusual rock, and is quite unlike any other rock unit in California, or elsewhere, apart from the closely associated Sisquoc Formation. It has very little sandstone, clay shale or conglomerate. Instead, it is a mixture of deep water organic oozes, volcanic ash beds and chemical precipitates such as silica. For the most part, it bleaches on weathering to a pale color ranging from white to tan. Its dearth of sand, silt and clay suggests that the shoreline lay so far from the deep basins that little land-derived material entered the basins which were, in some cases, more than 6000 feet deep.

During the time the Monterey Formation was being deposited, there was a great deal of volcanic activity in California and in the Santa Barbara area. This was likely the result of active subduction in the eastern Pacific sea floor under the continental margin. The abundant volcanic rocks found on the offshore islands and around Tranquillon Mountain in the southwestern Santa Barbara County are about the same age as the lower part of the Monterey Formation, showing that its deposition occurred during this volcanic activity. Among the characteristic features of the Monterey Formation are abundant, generally thin, volcanic ash beds. Many of these beds are less than an inch thick and are often not very obvious on casual inspection. They frequently

Figure 52.   Celite Quarry near Lompoc. Diatomite beds in the marine Miocene Sisquoc Formation.

promote landsliding because the ash alters to a slippery clay and where the formation has been tilted, blocks of rock can slide downhill on these greasy beds, particularly when they get wet and when lateral support is removed by erosion or grading (Figure 22). The ash beds also create weak zones and where the Monterey is present in the sea cliffs, marine erosion will work rapidly along the ash beds loosening blocks of rock and occasionally excavating small sea caves.

The large number of ash beds found in the Monterey Formation testifies to the frequency of volcanic activity and the composition of the ash shows that the volcanic activity was much the same type as we see today in the Cascade Range, rather than in Hawaii where volcanism is much less explosive and generally produces little ash.

Volcanic activity may have had a very significant effect on the abundance of diatoms in the sea at the time the Monterey Formation was being deposited. Diatoms, like all organisms, are limited by their food and mineral supply. The Monterey and the closely related Sisquoc formations are unusually rich in diatoms, so we know that conditions had to be very favorable for these organisms. One of the essential minerals required by diatoms is silica, from which they construct their frustules. Silica, however, is not very soluble in the sea, so the amount present is normally quite small, and may sometimes limit the abundance of diatoms in a given area. However, the frequent influx of volcanic ash, a material rich in silica, could well have provided any silica needed by the diatoms. Recently, it has become apparent that iron, too, may be a limiting factor for diatom growth. Volcanic activity also may have provided enough iron so that neither iron nor silica were limiting factors during Monterey and Sisquoc deposition.

Further, the depositional environment at that time appears to have been characterized by persistent upwelling, an oceanographic phenomenon in which winds and currents cause deep water to rise to the surface. The deep, upwelled waters typically carry more abundant nutrients than the warmer surface water.

The succeeding Sisquoc Formation shares many characteristics with the Monterey and grades into it in some places. However, in others the two are separated by a short erosional interval. The lower part of the Sisquoc is usually very well laminated, indicating that it was deposited in quiet, stagnant water which, like the Monterey, lacked bottom-living, burrowing organisms.

The commercial deposits of diatomite (also known as diatomaceous earth or *kieselguhr*) around Lompoc are of unusual purity and attest to quite a special sort of depositional environment in which only diatoms and the finest clay were being laid down in deep, quiet water (Figure 52). Diatoms, being one-celled green plants, live in lighted near-surface waters, so their abundance was not affected by the presence of stagnant water at depth. Similar conditions exist today in some of the deep basins in the Gulf of California and in parts of the Santa Barbara Channel.

The Monterey and Sisquoc formations are of major economic importance. The high organic content of these rocks has made them a major source of oil and gas in many parts of California. Not only are they generally believed to represent the source beds for much of the petroleum and natural gas found in our region, but because the Monterey also has been extensively fractured during later tectonic activity, it is also one of the main reservoir rocks. Fractures and bedding planes provide extensive pore space for fluids such as oil, gas and water. There are numerous places in the county where exposed Monterey rocks contain tarry oil in every fracture and exude a strong odor of petroleum. It is probable that the gradual buildup of heat and pressure associated with burial of the Monterey and Sisquoc organic-rich sediments converted some of the original organic materials into petroleum and natural gas.

Exposures of the Sisquoc Formation on the south coast and in the Purisima and Solomon hills have lower contents of organic material and more clay than do the high purity diatomites near Lompoc.

TERTIARY (PLIOCENE) AND QUATERNARY (PLEISTOCENE) ROCKS

The Pliocene Epoch was much shorter than the preceeding Miocene and lasted only from about 5 to 2 million years ago. The seas withdrew from most of the county except along today's immediate coast and in the Santa Barbara Channel, as well as in the Santa Maria area where an extensive marine embayment persisted until near the end of the Pliocene. The structural feature in which this embayment occurred seems to have formed as a result of plate movements between 17.5 and 6 million years ago (that is during the latter part of Miocene time).

During the Pliocene, the mountain-building period we call the Coast Range Orogeny began, and the Sierra Madre, San Rafael and Santa Ynez ranges were elevated and emerged from the sea. Likewise, the offshore islands were raised, though probably not to their present height. Even in the Santa Maria area where the large embayment persisted, some of today's hills like the Solomon, Casmalia, Purisima and Santa Rosa hills were folded into arch-like anticlines and began to emerge from the sea about 5 million years ago. A large area bounded on the south by the rising Santa Ynez Range, on the northeast by the Little Pine and related faults and the Sisquoc River, and on the west by the sea, remained under water in which the fossiliferous Careaga sand was deposited.

This marine embayment is known as the Santa Maria Basin and today the northern part of it is known as the Santa Maria Valley. In addition, the modern coastal plain from Goleta, eastward to Carpinteria, remained largely under the sea during the Pliocene, continuing into the Pleistocene when uplift produced a flight of elevated marine terraces along both the southern and western coasts of the county.

These uplifted benches are especially prominent in the Santa Barbara city area (Figure 53) and much of the urban development has taken place on them. As ages of these terraces have been determined, it has become apparent that terrace elevation is no guide to terrace age, although within a group of terraces in one location, the oldest would certainly be the highest. In Isla Vista one terrace, dated at 42,000 years old, has an elevation of only about 40 feet. To the east, at More Mesa and Hope Ranch, the terrace of the same age is several hundred feet high. If any evidence were needed to show that tectonic activity is still in progress in Santa Barbara County, this would provide it. Further, east of Santa Barbara, some Pleistocene rocks, originally horizontal,

Figure 53. More Mesa near Goleta. The elevated marine terrace, cut about 40,000 years ago, has been uplifted about 85 feet above sea level. The northern edge of the terrace lies along the More Ranch fault and Atascadero Creek. Exposed in the sea cliff are marine Plio-Pleistocene rocks of the Santa Barbara Formation.

have been tilted 60° or more from their original position. Other evidences of continued tectonic activity include historic records of earthquakes and faults in which Plio-Pleistocene marine rocks have been uplifted out of the sea and brought into contact with older rocks. Further, continuing creep has been measured along some of the county's faults.

Deposits of Pleistocene age typically are quite often weakly cemented and barely deserve the name "rock" - they are perhaps better considered rocks-in-the making. For the most part, they may not have existed long enough for their materials to be cemented or for compression to form coherent rock. Conversion of loose sediment into consolidated rock is called **lithification**.

Careaga Formation

The most widespread Pliocene rock type in the county is the Careaga sandstone found mainly north of the Santa Ynez River, particularly in the Purisima and Solomon hills. Most of this rock unit was deposited in a very shallow marine basin, as demonstrated by the abundant presence of sand dollars that occur in it on the north flank of the Purisima Hills.

Santa Barbara Formation

On the south coast, in the Santa Barbara-Goleta area, is another shallow water marine deposit, partly of Pliocene age. This is the Santa Barbara Formation, but unlike the Careaga, much of this rock is of Pleistocene age showing that marine conditions persisted longer on the south coast than they did in the Purisima and Solomon hills.

The Santa Barbara Formation is mostly poorly consolidated and easily dug with a shovel. For many years it was quarried for fill sand at the west end of Valerio Street and in the Las Positas Park area. Like the Careaga Formation, it is richly fossiliferous in some places. One such area, now largely eliminated by harbor development is at the former location of Point Castillo, where the Santa Barbara breakwater reached the shore. This locality was long known to paleontologists as "Bath House Beach". A small remnant of this fossiliferous rock is present in the hillside at the east end of the Santa Barbara City College football field.

The Santa Barbara Formation also is exposed on the east and north sides of the hills from the harbor area to La Cumbre Middle School and as far west as Hope Ranch. Another exposure is north of U.S. Highway 101 in the vicinity of the sheriff's office and rubbish transfer station near El Sueño Road.

In the Goleta area, the formation is largely buried beneath younger alluvial deposits, but is exposed in the cliffs at More Mesa (Figure 53), and at several places in the lower foothills of the Santa Ynez Mountains. The Santa Barbara Formation is a very important aquifer in the Santa Barbara and Goleta areas. Most of the water wells in this area pump from this formation.

Casitas Formation

Younger deposits of Pleistocene age include the gravelly Casitas Formation exposed in the cliffs at the west end of Summerland Beach, at Loon Point east of Summerland (Figure 30) and in the hills on either side of Rincon Creek near its junction with Casitas Creek. The Casitas Formation is a non-marine, land-laid deposit.

Paso Robles Formation

The age-equivalent of the Casitas Formation north of the Santa Ynez Mountains is the very widespread Paso Robles Formation. This older alluvial deposit flanking the San Rafael Mountains and composed to a large extent of debris from the Monterey Formation is exposed in the nearby uplands. It is full of bleached whitish chips of Monterey shales and forms most of the low, rolling hills that surround the San Rafael Mountains, as well as those that flank the northern side of the Santa Ynez Range. The broad flats that lie above the stream valleys in both the Los Alamos and Santa Ynez valleys are mostly on the Paso Robles.

The Fanglomerate

Youngest of the Pleistocene deposits is the Fanglomerate, sometimes called the older alluvium. It is present on both sides of the Santa Ynez Mountains and along the flanks of the San Rafael Mountains in the Santa Ynez Valley area. In the Santa Barbara city area, it is present on Mission Ridge from the vicinity of Sycamore Canyon westward to the Santa Barbara Mission, around Sheffield Reservoir and the San Roque area on either side of Foothill Road, and at the Botanic Garden.

As noted in the discussion of the Coldwater and Matilija sandstones, the most interesting feature of this formation is the enormous sandstone boulders it contains, many of which are up to 8 feet across. These boulders are so large that no flood in historic times has been capable of moving them. The agent responsible was almost certainly mudflows during storms of almost unimaginable intensity during the Pleistocene glacial stages when the local climate was much wetter and colder than at present. Some of the larger prehistoric landslides in southern California are of late Pleistocene age and may also be attributable to unusually intense storms at that time.

CHAPTER 4

STRUCTURAL FEATURES

Santa Barbara County is in the area where the southern Coast Ranges, with their northwesterly trend, merge with the east-west trending Transverse Ranges (Figure 2 and 3). The Channel Islands, the Santa Barbara Channel and the Santa Ynez Range are all dominated by structural features that trend east-west, and are an integral part of the Transverse Range Province of California. In contrast, the western part of the county north of the Santa Ynez fault and Santa Ynez Mountains is characterized by faults and folds that more or less parallel the great San Andreas fault which lies only 4 miles beyond the northeastern corner of the county in the Cuyama Valley area.

The eastern part of the county north of the Santa Ynez Range could be included, structurally speaking, in either the southern Coast Ranges or in the western Transverse Ranges—it is truly a transitional area. For example, the eastern part of the Big Pine fault trends east-west, matching the trend of the Transverse Ranges, but on the west, the Big Pine fault curves around to match the general trend of the Coast Ranges (Figure 2).

This is yet another example of nature's disdain for man-made distinctions and classifications. Nature almost always deals in gradations rather than in well-defined boundaries.

FAULTS

Each of the major faults in the county will be described from south to north, beginning with the islands, and concluding with the Cuyama River area. In urban areas, some of the lesser faults will be considered because they have prominent topographic effects or because they have a record of activity or pose a threat of one. Figure 2 shows the location of the major faults in the county.

Faults are fractures in the rocks along which some movement has occurred. There is a myriad of fractures in rocks along which no displacement has occurred and these are called "joints". Most people are aware that faults have something to do with earthquakes because the news media often report that a given earthquake was produced by this or that fault. Some earthquakes, to be sure, are produced by processes such as volcanic activity, but most are the result of abrupt displacement of rocks along a fault.

Three main sorts of displacement can occur along faults and these often merge with one another. The first of these is essentially horizontal offset along a fault, where neither block is raised or depressed with respect to the adjoining block. Most of the major faults in Santa Barbara County are of this type. If these faults are traced downward into the crust, they are nearly vertical fractures. Such faults are called "strike-slip faults" and the San Andreas is the classic example of this form.

The second category is the result of crustal stretching in which some blocks drop downward like keystones with respect to adjacent blocks. The fault plane may be either vertical or inclined, and if inclined, these faults are called "normal" faults, perhaps an unfortunate name because other types of faults are no less normal.

The last category are those faults that result from crustal shortening or squeezing from both sides. These are called "thrust" or "reverse" faults in which one block rides up and over the adjacent block. In the Transverse Ranges, such faults commonly form the boundaries of mountains like the Santa Ynez Range. The Santa Ynez Range, the Santa Barbara Channel and the offshore islands have been squeezed from north to south, crumpled into folds and broken by faults. Perhaps the most dramatic example of this crustal shortening in the county is the Montecito overturn, on the south face of the Santa Ynez Range, where folding has been so intense that some rocks are now upside down (Figure 5). Wherever older sedimentary rocks rest on younger sedimentary rocks, reverse faulting or intense folding will be the cause.

Nature, however, seldom puts faults into such neat boxes and in a great many cases a given fault will be found to have both a vertical as well as horizontal component of displacement. This may be clearer if we consider the San Andreas fault in California. Where it has a northwesterly trend, its displacement is chiefly horizontal and right-slip which means that an observer standing on one side of the fault, looking across it, would see that the block on the far side had been shifted to his right. Where such a fault veers left as it does crossing the Transverse Ranges, the blocks on opposite sides are shoved *into* one another as well as shoved *past* one another. This squeezes up mountains like the Transverse Ranges and many of the lesser faults resulting from such squeezing are reverse or thrust faults.

Santa Cruz Island and Santa Rosa Island Faults

It has been suggested that these two faults whose connection, if any, is concealed beneath the channel separating the islands, are parts of a much larger, east-west fracture system that begins at least as far west as San Miguel Island and extends eastward to the mainland and along the southern side of the Santa Monica and San Gabriel mountains in Ventura and Los Angeles counties. The underwater connections are still in dispute. In any event, the Santa Cruz Island fault is clearly a left-slip fault with the south side up. This is obvious because the Mesozoic igneous and metamorphic rocks on the south side of the fault are in contact with the much younger Tertiary volcanic rocks on the north side and because stream courses crossing the fault have been offset to the left.

The Santa Cruz Island fault accounts for the development of the unique, elongate Central Valley of the island. The fault encouraged erosion in the crushed and broken rocks along the fault zone (Figure 10).

Although we have no historical record of earthquakes generated by movement along this fault, it is long enough and young enough to pose a potential risk. The offsets in the small stream channels crossing the fault are regarded as indicative of geologically young movement.

Aggregate displacement on the Santa Cruz Island fault has been estimated to be about 11 miles horizontally and 7500 feet vertically.

The Santa Rosa Island fault is quite similar to the Santa Cruz Island fault, though erosion along it has failed to produce a valley. The nature of slip on this fault as well as its magnitude is much the same as for the Santa Cruz Island fault, though the amount of vertical offset appears to be less, but the left offset of streams is more prominent on Santa Rosa Island than on Santa Cruz.

Santa Barbara Channel Structural Features

Oil drilling and exploration based on many years of geophysical investigation, show that the rocks on the channel floor have been deformed into several well-defined east-west trending belts of folds and faults. Some of these features are many miles long and some of the faults may pose appreciable earthquake risk.

The two main structural trends are known as the Rincon and Montalvo trends. The Rincon trend joins the Red Mountain fault zone in Ventura County at Punta Gorda. From this point, the trend is easily seen from shore as most of the oil drilling platforms are along this feature. The trend continues inshore in Ventura County and is marked by a string of productive oil fields to the east.

The Montalvo trend lies farther offshore, approximately half way between the islands and the mainland. There are fewer oil platforms on this trend, in part because it lies in deeper water.

Carpinteria Fault

This relatively short fault is exposed in the big road cut on the east side of U.S. Highway 101 between Rincon Point and Carpinteria. In years past, on the elevated marine terrace seaward of the highway and just west of it, was an auto race track called the "Thunderbowl" (Figure 28). The slight depression occupied by the Thunderbowl is a sag pond on the Carpinteria fault produced by tension that acted to pull apart an area between two offset strands of the fault (Figure 54).

The Carpinteria fault goes out to sea at Sand Point at the mouth of the salt marsh, producing a mostly submerged rocky reef that is exposed only at the very lowest tides a few times each year. The importance of this reef was discussed in the earlier section on headlands.

Arroyo Parida-Mission Ridge-More Ranch Fault Zone

This fault zone begins in Ventura County near Ojai and extends westward into Santa Barbara County at a point about 1.5 miles inland from where State Highway 150 crosses Rincon Creek. At this point, the fault is known as the Arroyo Parida fault. It disappears beneath alluvial deposits in Picay Creek north of Summerland, and reappears as the Mission Ridge fault along the north-facing side of the Riviera district of the city of Santa Barbara, passing just south of Sheffield Reservoir and along the low hills in the Loreto Plaza area of upper State Street. It appears to cross State Street near where that street and De La Vina join. The Mission Ridge fault joins the Mesa fault near the Five Points Shopping Center and extends westward along the north side of the Hope Ranch Hills, close to Atascadero and Cieneguitas creeks. This western segment is known as the More Ranch fault. It crosses Mescalitan Island

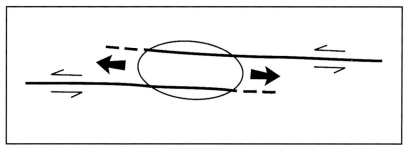

Figure 54. Diagram of the pull-apart structure along the Carpinteria fault that formed a sag. This structure was the site of the former Carpinteria "Thunderbowl". (See also Figure 28).

in the mouth of Goleta Slough, follows the north side of the UCSB campus, finally going out to sea at the Sandpiper Golf Course southwest of the Ellwood area. The south side of this fault has been elevated several hundred feet with respect to the north side.

There is little doubt that this fault system could generate earthquake activity in the Santa Barbara area, but on the other hand, it forms a valuable impervious subsurface barrier that protects the ground water in the Goleta Valley from sea water intrusion. Many wells on the north side of the fault pump potable water from below sea level, mostly from the very porous and permeable sands of the Plio-Pleistocene Santa Barbara Formation (Figure 5).

Some years ago, a proposal was made to dredge part of the Goleta Slough to create a small boat harbor. It is probably fortunate that this plan did not eventuate because dredging of the clays in the Slough might have destroyed the seal that prevents sea water from moving downward along the More Ranch fault into the fresh water aquifer that lies beneath the Goleta Valley.

Mesa Fault

The Mesa fault extends from the Five Points Shopping Center southeasterly to the tip of Stearns Wharf. It goes under the Carrillo Street U.S. Highway 101 underpass where persistent seeps occur, though it is nowhere well exposed. Neither the Mission Ridge nor the More Ranch faults have good exposures. Their locations, like the Mesa fault are determined mostly by topography and well records, springs and seeps (Figure 9).

Although any member of this group of faults could generate a damaging earthquake, so far only the Mesa fault seems to be guilty. The destructive 1925 earthquake has been attributed to movement on the offshore portion of this fault.

There are a number of other named and unnamed faults in the Santa Barbara city area. Many of these extend northwesterly into the Santa Ynez Mountains where they die out. Some displace Quaternary deposits so young that the faults must be considered potentially active. Faulting is so pervasive in the Santa Barbara-Goleta area that few locations are much more than two miles from a potentially active fault and it is well to remember that one's house does not need to be on top of a fault to experience considerable shaking or damage during an earthquake.

Santa Ynez Fault

The high, single-crested Santa Ynez Range was elevated mostly during Quaternary time along the Santa Ynez fault. This fault should be considered active. It extends westward from Ventura County to Gaviota Pass on U.S. Highway 101 where it divides into two branches, the Pacifico fault on the north that dies out as an anticlinal fold in Jalama Canyon, and the south branch that cuts across U.S. Highway 101 just north of the Gaviota Pass tunnel and goes out to sea just west of the mouth of Alegria Canyon on the Hollister Ranch.

Displacement on the Santa Ynez fault is such that the south side is up relative to the north side. It has been asserted that this fault was originally a more gently inclined thrust fault along which rocks of the present Santa Ynez Range were shoved northward. According to this view, the fault has been steepened by continued crustal shortening in a north-south direction crumpling and folding the rocks more and more intensively, eventually overturning some of those on the seaward side of the mountains from Montecito eastward to the county line. Even greater horizontal displacement on the fault has been proposed, based on offset stream courses, marked differences in the sequences of rocks on either side of the fault and by other structural features. As much as 20 miles of left slip has been proposed, though there remains considerable uncertainty about this amount.

Franciscan rocks are not exposed south of the Santa Ynez fault except at Blue Canyon and Forbush Flat north of Montecito, although they may be deeply buried elsewhere in the Santa Ynez Mountains.

The Santa Ynez fault forms the boundary of the high part of the Santa Ynez Mountains and parallels the trend of the Santa Ynez River. As noted earlier, north of this fault most of the structural features show the typical northwesterly trend of the Coast Ranges, hence in Santa Barbara County, the Santa Ynez fault marks a distinct boundary between the Transverse Ranges to the south and the Coast Ranges to the north. The boundary is much less distinct or sharp in Ventura County to the east.

Santa Ynez River Fault

Several geologists have suggested yet another branch of the Santa Ynez fault, diverging near Lake Cachuma and continuing a bit north of west to the coast west of Lompoc. This structural feature, if indeed it exists, is nowhere exposed and its existence is based on the presence of a sequence of sedimentary rocks from Late Cretaceous to Early Miocene age which are present south of the fault, but absent to the north. Other geologists are very skeptical and offer other explanations for the differing rock sequences on either side of the Santa Ynez fault.

Little Pine Fault

This is a very prominent southern Coast Range fault that terminates against the east-west trending Santa Ynez fault near the upper end of Gibraltar Reservoir and roughly parallels other Coast Range structural features that have the typical northwesterly trend. The Little Pine fault is clearly traceable to a point about 1.5 miles east of where Zaca Lake and Foxen Canyon roads intersect. Beyond this point, it either dies out or may merge with the Foxen Canyon fault that can be traced northwesterly as far as the Sisquoc River.

In many places the trace of this fault is strikingly evident because it forms the boundary between the very distinctive Franciscan suite of rocks on the northeast—serpentinites, greenstone, blueschist, dark graywacke sandstone and reddish bedded chert that contrast strongly with Tertiary sedimentary rocks on the southwest. In addition, the Tertiary rocks form more rounded and subdued topographic features compared to the knobby, hummocky and more irregular exposures of Franciscan rocks (Figure 55).

The Little Pine fault and its segments are thrust faults along which the older Franciscan rocks have ridden up and over the younger Tertiary rocks on the southwest. When fossils or other evidence tells a geologist that older rocks rest on younger rocks, there are two usual explanations: (1) A thrust fault may be involved as is the case along the Little Pine fault, or (2) folding may have been so intense that the normal order of rocks has been reversed by overturning, as is the case on the south flank of the Santa Ynez Mountains eastward from Montecito. Which of these explanations is the correct one usually becomes evident with detailed mapping.

Camuesa Fault

This fault can be followed from lower Indian Creek north of Gibraltar Reservoir northwesterly to Figueroa Mountain. Along this fault, the Franciscan rocks on the southwestern side are in contact with younger sedimentary rocks on the northeast. This displacement can be explained either by uplift of the rocks on the southwest side of the fault, or by horizontal displacement of these same rocks with respect to those on the northeast side of the fault. The fault appears to be nearly vertical.

Big Pine Fault

The Big Pine fault is one of the major structural features of southern California and can be traced from Cuddy Valley in southernmost Kern County where it abuts against the great San Andreas fault, southwesterly through Ozena on State Highway 33 in Ventura County into Santa Barbara County south of Lizard Head to Bluff Camp on the south face of Big Pine Mountain (the County's highest peak, 6826 ft). The fault then curves around the south side of Santa Cruz Peak and swings northwesterly to parallel southern Coast Range structural trends, dying out against the Camuesa fault in upper Cachuma Canyon (Figure 2).

The Big Pine fault is a left slip fault, as is the large Garlock fault in the desert area of eastern California. Because these two large left slip faults end on opposite sides of the San Andreas fault, but are offset from one another by about 6 miles, some geologists have suggested that they may be parts of a single structure, offset by later movement along the San Andreas fault.

On the basis of some vague linear features suggested by high-altitude photography, a westward extension of the Big Pine fault from near Figueroa Mountain to Point Sal has been proposed, but such an extension is neither evident from field studies nor from subsurface data. In addition, it would require offset of other structures such as the Camuesa and Little Pine faults.

Figure 55. Figueroa Mountain; view north. The pine-covered upland at the crest of the mountain lies on synclinally folded Monterey Formation. The hummocky grass-covered area is underlain by the Franciscan Formation, bounded on the uphill side by the Camuesa fault and on the lower side by the Little Pine fault. The smoother grass-covered hills in the foreground are developed on the nonmarine Paso Robles Formation.

In 1852, there was a strong earthquake causing ground breakage in Lockwood Valley. It is generally assumed that the Lockwood Valley in question was the one in northern Ventura County through which the Big Pine fault passes. However, there is also a Lockwood Valley in southern Monterey County and this ground breakage may have occurred there on a totally different fault. The fragmentary records of this earthquake simply do not allow us to be certain of the exact locality. Unless the Big Pine fault obliges us with a new earthquake somewhere along its trace, we may never solve this problem. Nevertheless, most of the movement on the Big Pine fault appears to be younger than 2 million years and to total between 8½ and 18 miles.

Rinconada Fault

Older maps show this fault as part of the Nacimiento or Sur-Nacimiento fault, but more detailed mapping has now demonstrated that the two faults are distinct, but overlapping near King City in Monterey County. The Rinconada fault is a major feature about 160 miles long, and for much of its length it forms a very important crustal boundary.

The enigmatic crustal structure called the Salinian Block (Figure 43) is bounded in Santa Barbara County by the Rinconada fault on the southwest. Salinian granitic basement presumably underlies the Sierra Madre and Cuyama Valley northeast of the Rinconada fault. This fault was most active during the Pliocene Epoch when the Franciscan rocks on the southwest moved northwesterly, possibly as much as 120 miles. This fault is part of the northwesterly-trending San Andreas family of faults, and like the San Andreas it has right slip movement.

The trace of the Rinconada fault across Santa Barbara County begins at the Big Pine fault near Dutch Oven Campground south of Madulce Peak in the upper Indian Creek drainage. The fault strikes northwest and is partly followed by the upper Sisquoc River from north of Big Pine Mountain to Mine Canyon south of McPherson Peak in the Sierra Madre Mountains. From there, it parallels the trend of the Sierra Madre but lies about 3 miles southwest of the range crest. It passes through Pine Flat near Miranda Pine Mountain and crosses the Cuyama River into San Luis Obispo County near Clear Creek. For a large fault, it is generally quite poorly exposed, and unlike the San Andreas is not marked by prominent furrow, offset streams or other features that one might expect along such a large, right-slip fault. The poor exposures very likely result because the Rinconada has had little, if any, activity during Quaternary time (the past 1.8 million years), so erosion has had a chance to erase most topographic evidences that may once have been present.

The Rinconada fault seems to be nearly vertical and probably has some vertical offset in addition to its dominant horizontal displacement. The best evidence for vertical displacement is the elevated Sierra Madre block which lies between the Rinconada and Ozena faults.

Ozena and South Cuyama Faults

This fault system, like the Rinconada, terminates against the Big Pine fault in Ventura County near the head of the Cuyama River at the north base of Pine Mountain. The upper Cuyama River in Ventura County roughly follows the trace of the Ozena fault, but in Santa Barbara County, the Ozena fault cuts across the northeastern flank of the Sierra Madre. Northwest of Branch Canyon, however, the poorly exposed South Cuyama fault defines the northeastern base of the Sierra Madre Mountains.

Although much of the displacement on these faults is probably right slip like the nearby San Andreas, the Ozena fault shows thrust offset where it crosses Santa Barbara Canyon near the junction with Rhoda Canyon just east of Fox Mountain. The fault dips under the Sierra Madre and the central part of that range has been shoved up and over younger rocks to the northeast on the Cuyama Valley side.

FOLDS

To the geologist, the term "folding" refers to the bending of rock strata without breaking. In order for brittle materials like rocks to bend, they must either be quite soft and plastic, or more likely be under high confining pressure. This means, of course, that the folding we see in surface rocks was either developed when the rocks were still soft and unlithified, or that it took place when the rocks were deeply buried beneath a thick overburden.

Movements of the earth's large crustal plates stress the rocks which are then deformed by both folding and faulting. When stresses exceed rock elasticity, the rocks break and a fault is formed the moment any offset occurs along the break.

Folding can tell us a good deal about the timing of periods of mountain building. It was earlier noted that the mere presence of high, steep mountains shows that the forces of deformation and uplift have more than kept pace with the destructive forces of erosion which act to erase all highland areas. When we look at the deformed rocks that make up our mountains and determine by fossils or other means when they were deposited, we can bracket the time of deformation or mountain building. If we find marine Pleistocene rocks that originally were deposited as flat or nearly flat sheets, but are now inclined 50° to 60° or even more, as we do near Summerland and Carpinteria, we know that crustal deformation has been active in that area during the past 1.8 million years (very young by geological standards). The Casitas Formation exposed in the seacliff at Loon Point near Summerland is such an example (Figure 30). Near Lavigia Hill, overlooking Santa Barbara's La Mesa area, marine sedimentary rock of the Quaternary Santa Barbara Formation now dips 30° in some places.

Deformation can be shown by simple uplift without much folding or faulting. Elevated marine terraces are common along the Santa Barbara coastline (Figure 52). Nearly all these terraces are benches cut by the sea in the surf zone and are of Pleistocene age. Some between Carpinteria and Rincon Point are now 240 feet above sea level. Because the formation of the high terrace has been dated, we know that during the past 42,000 years it has been raised about 240 feet near Rincon Point, depressed below sea level west of Carpinteria, and raised more than 600 feet near Punta Gorda in Ventura County only 4 miles southeast of Rincon Point. This same 42,000 year old terrace lies about 240 feet above sea level at Hope Ranch west of Santa Barbara and only about 40 feet above sea level on the UCSB campus at Goleta. Not only does this show that deformation is very recent, indeed ongoing, but also that one cannot assume a terrace now at an elevation of 100 feet, for example, is necessarily the same age as another terrace at that elevation a few miles away. One cannot determine terrace ages based on elevation alone.

This deformation seen along the terraces usually called **warping** to distinguish it from **folding** that produces anticlines and synclines, more or less at right angles to the strike or trend of rock layers. This distinction may seem too subtle for most readers, so be assured that both types of deformation can properly be considered **folding**. All the hills and mountains include folded rocks (Figures 5-7). As has been noted, folding in the Santa Ynez Range has been so intense east of Montecito that some of the rocks have been completely overturned and are now upside down with respect to their original depositional position. See, particularly, the contact between the Cozy Dell and Coldwater formations in Figure 5.

Although it is certainly true that all the higher ranges in the county are at least in part due to elevation by faulting, some of the lower hilly areas such as the Purisima and Solomon hills are entirely folded features in which faulting has played no direct role at all.

In the chapter where the nature of the county's hills and mountains was considered, you may have noticed that many of the high peaks in the San Rafael Mountains are located on synclinal folds. In contrast, the lower hilly areas are more likely to be anticlinal folds with synclinal folds under the intervening valleys. For example, Los Alamos Valley is synclinal and the Purisima and Solomon hills on either side are anticlinal.

What might account for this difference?

When folding begins to affect an area, the surface of the ground will be bowed upward over anticlines and downward over synclines. As folding continues, the anticlines usually become more tightly folded and will certainly be elevated as well. As was previously pointed out, the rocks in the crest of the anticline are stretched and fractured while they are being elevated. Both these things concentrate erosional processes along the anticlinal crest, but will have pretty much the opposite effect on the adjacent synclines. Eventually, as the area continues to be elevated, the original topography is reversed, leaving synclinal folds at higher elevation than the anticlines.

EARTHQUAKES

From the time the Spanish first began to settle southern California in the late 1700's, and wrote down their impressions of the land, we get reports of earthquakes. Historical records of earthquakes span just a little more than 200 years in our area and earthquakes have been documented with recording instruments for only a little more than 100 years. During this time we have learned a great deal about earthquake origin and propagation and about the potential for building damage, but there are large uncertainties about the frequency of major earthquakes and we have yet to do very well in predicting future quakes.

Mission records tell us that each of the three Spanish Missions in Santa Barbara County – Santa Barbara, Santa Ines, and La Purisima Concepcion – have sustained repeated, sometimes very severe, damage. The great earthquake of 1812, likely originating on a sea-floor fault in the western Santa Barbara Channel, totally destroyed Mission La Purisima Concepcion only 10 years after it had been built. The Mission was later rebuilt in another, supposedly safer location. This same earthquake destroyed the chapel at Mission Santa Ines and Santa Barbara Mission also had to be rebuilt. Similarly, the Fort Tejon earthquake of 1857, perhaps the strongest earthquake in the recorded history of southern California, did considerable damage at Santa Barbara Mission.

From July to December 1902, a swarm of earthquakes struck Los Alamos in the central part of the county, almost totally destroying buildings in the small town. Neither the epicenter nor the magnitude has been established for these quakes, but many settlers fled the area and Los Alamos was mostly abandoned for years.

The magnitude 6.3 Santa Barbara earthquake of 29 June 1925, evidently originated offshore, possibly on the seaward extension of the Mesa fault. Over 130 buildings in the central business district were either destroyed or damaged

Figure 56. Santa Barbara following the devastating earthquake that occurred on 29 June 1925. Note the extensive damage to buildings in the 900 block of State Street. (Photographer unknown).

so badly they could not be used until repaired (Figure 56). Mission Santa Barbara sustained considerable damage and the bell towers had to be replaced. To date, the Santa Barbara Mission has sustained damage in six or eight earthquakes that have affected the area since 1780.

For the most part, earthquake damage has occurred along the south coastal part of the county, mostly, of course, because this is the area of longest urban development, but it is also possible that there are more active faults in that part of the county as well as offshore in the channel.

The 1925 earthquake killed 13 people and caused perhaps as much as $20 million in damage. It is generally agreed that in the long term, it was a blessing for the city, because it did so much damage in the central business district. Many buildings had to be replaced or extensively repaired, and local leaders took the opportunity to change a dowdy, undistinguished main street into an attractive blend of California-Mediterranean architecture. As a result, Santa Barbara became one of the more attractive cities in California which has attracted millions of dollars from visitors and vacationers in the decades following the earthquake. Oddly, exactly one year after the 1925 quake, on 29 June 1926, another strong earthquake shook Santa Barbara and resulted in further damage and one death.

On 4 November 1927, a very strong earthquake (magnitude 7.5) occurred on a submarine fault off Point Arguello. This quake generated a small tsunami at Point Arguello that reached heights between 5 to 7 feet. Shaking from this earthquake resulted in some building damage at both Lompoc and Santa Barbara.

On 30 June 1941, Santa Barbara was again shaken by a strong earthquake (magnitude 5.9), evidently also originating offshore, again possibly on the offshore extension of the Mesa fault. Moderate building damage occurred in the city.

The Kern County earthquake of 21 July 1952 (magnitude 7.7), was the strongest earthquake to affect California in the 20th century except for the San Francisco earthquake of 18 April 1906. Although the epicenter was near Arvin, not far from Bakersfield, Santa Barbara was strongly shaken and about $400,000 of damage was done to homes and commercial buildings in the city. The Carrillo Hotel in the central business district experienced broken windows and damaged brick work. The night before the earthquake (which happened about 6 a.m.), one of the bellboys on duty had helped a guest get into bed after considerable overindulgence at some local watering holes. This guest slept through the earthquake and when he awoke about 10 or 11 in the morning, he was horrified to see furniture in his room overturned and other signs of disarray. He rang the desk and began to apologize for what he thought was his fault, only to be cheerfully told, "Oh! You should see what it looks like down here!"

During the 1952 earthquake the Tecolote water tunnel under the Santa Ynez Range was under construction and the tunnelers were working around the clock. Their families and others were concerned that these men might be trapped by falling rock, only to discover that when the tunnelers came off shift they had noticed scarcely any shaking. It has long been known that the amount of shaking during an earthquake is much less evident on bedrock than on the alluvial deposits where most people live. Filled marshy areas such as those at Santa Barbara Airport and on the lower east side of the city are subject to the most severe shaking.

On 4 July 1968, the strongest of a swarm of earthquakes occurring over a six-week period struck the Goleta area. This earthquake originated on a sea floor fault midway between Santa Barbara and Santa Cruz Island. Santa

Barbara city had virtually no damage, suggesting the existence of some sort of focusing mechanism, a phenomenon clearly evident some ten years later on 13 August 1978, at 4 in the afternoon. This strong (magnitude 5.9) earthquake, like so many others, originated on an offshore fault and like the 1968 earthquake, was focused toward Goleta Point and the UCSB campus. Close to a million books were unshelved in the library, office bookshelves toppled and plaster and fluorescent lamp fixtures were damaged in many buildings. The entire Marine Science building was found to have shifted more than an inch. The cost of these effects and other structural damaged amounted to about $3.5 million. There were no injuries or deaths, in part because the quake occurred on a Sunday afternoon in the summer when few students or staff were on the campus. However, near Ellwood, an hour or so later, a train was derailed when it encountered a place where the rails had been spread by the quake.

While the foregoing represents a summary of the more important earthquakes to affect the county since about 1780, it is by no means a complete list and numerous other earthquakes, some quite strong, have occurred and have been noted in various historical records.

CHAPTER 5

GEOLOGIC SKETCHES OF THE OFFSHORE ISLANDS

Santa Barbara Island

This is the smallest of the eight islands off southern California and has an area of only about one square mile. It is a part of Channel Islands National Park and has limited visitor facilities (Figure 57). The island has two prominent offshore islets, Sutil on the south and Shag Rock on the north. These islets and Santa Barbara Island itself are almost completely surrounded by cliffs 200 to 600 feet high. As a result, the island has no beaches worthy of the name. Much of the island is flat-topped due to Pleistocene marine and terrestrial erosion that has cut a flight of six recognizable marine terraces or benches, the highest of which is about 250 feet above sea level. Some of the terraces have thin, fossiliferous marine sediments capping them which allows us to date their formation.

Bedrock consists of a thickness of about 1000 feet of tilted pillow lavas of Miocene age best exposed in the high cliffs. The pillow structure shows that the lava flows were erupted under water, as does a thin, discontinuous layer of middle Miocene mudstone capping the pillow lavas. Figure 37 illustrates what pillow lavas look like, albeit from another locality. Above the mudstone is several hundred feet of rubbly blocks of lava mixed with limy marine deposits, showing that this rubbly rock, called an agglomerate, was also formed under water.

Uplift probably began in the Pliocene about 5 million years ago, and continued into the Pleistocene as evidenced by the terraces. However, the Pleistocene was also a time when sea levels rose and fell 300 feet or more as the great continental glaciers waxed and waned. During the low sea levels when the ice sheets were at their maximum size, the island was considerably larger than at present. The post-glacial rise in sea level and wave erosion are steadily reducing the island's modest size.

Figure 57. Sea Elephant Cove on the north coast of Santa Barbara Island. All the rocks in the view are Miocene volcanic rocks, mostly basalt and andesite breccias, and pillow lavas. Nearly all appear to have been erupted under water.

Santa Cruz Island

This is the largest of all the offshore islands in southern California and in many ways the most varied geologically and topographically. It is also the highest of the eight California islands, reaching 2167 feet at Picacho Diablo in the western part of the island.

As previously remarked, Santa Cruz Island is bisected longitudinally by the Santa Cruz Island fault zone along which the prominent central valley, Cañada del Medio, has been eroded (Figure 10). The greater part of the central valley drains to Prisoners Harbor on the north coast via a winding gorge about 2 miles long called Cañada del Puerto. The extreme eastern end of the valley drains easterly and the western end drains into the broad, west-facing bight at Christi Ranch.

North of the fault and along the central valley west of Cañada del Puerto, the rocks are almost entirely basalts and andesites, mostly in the form of lava flows and breccias, evidently chiefly deposited on land. While these volcanic rocks resemble those in the western part of the Santa Monica Mountains on the mainland in Ventura and Los Angeles counties, they have a somewhat different mineralogy and were likely derived from different eruptive centers.

Chapter 5   115

Figure 58.   Potato Harbor at the northeastern end of Santa Cruz Island; easterly view. The light-colored rocks in the cliffs belong to the marine Pleistocene Potato Harbor Formation. The darker rocks are part of the Santa Cruz Island volcanics of Miocene age.

Perhaps surprisingly, these volcanic rocks do not occur anywhere on the south side of the Santa Cruz Island fault. Different rock types of middle Miocene age are in contact with one another along this fault. These relationships and some other evidence indicate that there has been considerable left slip on this fault as well as probable vertical offset.

Eastward from Chinese Harbor on the northern coast, the volcanic rocks are partly covered by the marine Monterey Formation which extends southeastward across the island at its narrowest point. The extreme eastern part of the island beyond a prominent ridge called El Montañon, is mostly developed on volcanic rocks except at Potato Harbor where there is an exposure of Plio-Pleistocene marine rocks in the sea cliff, similar to the mainland Santa Barbara Formation (Figure 58).

South of the Santa Cruz Island fault, rocks are distinctly different from those north of the fault. Oldest on the south side are the schists, intrusive diorites and related rocks. These form a band almost 10 miles long on the south side of the fault from where it goes out to sea on the east, westward almost to the Christi Ranch.

In the southwestern part of the island, underlying the Tertiary volcanic rock is a thick pile of sedimentary rocks ranging in age from Paleocene (about 50 million years old) through Eocene, Oligocene and Miocene. The total thickness of this sedimentary sequence is about a mile. Best exposures of these rocks are in the Sierra Blanca area south of the Christi Ranch and inland from Kinton Point near Pozo Creek.

Middle Miocene volcanic rocks more or less encircle the old crystalline basement (the schists and diorites), usually separated from one another by faults, but in a few places the volcanic rocks rest directly on eroded surfaces developed on the older rocks.

Some of the more interesting sedimentary rocks in this part of the island are the pebble conglomerates of the Eocene Jolla Vieja Formation, the very unusual breccias of the Miocene San Onofre Breccia and the pale sandy, tuffaceous volcanic sandstone of the Blanca Formation (Figure 59).

The reason the pebbly Jolla Vieja Formation is of interest is because very similar pebbly rocks of like age occur on San Nicolas, San Miguel and Santa Rosa islands and near Poway in the San Diego area. These pebbles and cobbles are rocks quite unlike anything elsewhere in the coastal region of southern California. Where did they come from, and how did they get such a disjointed distribution? The most likely source so far identified is in northwestern Sonora in Mexico. The present distribution is most likely due to the complex horizontal and rotational movements among a host of southern California faults and crustal blocks, providing good evidence that southern California geography has changed greatly since Eocene time when the Jolla Vieja Formation and its corresponding counterparts were deposited.

The San Onofre Breccia represents an exposure of one of the more intriguing and unusual rocks of southern California. It was first described in 1925 from an exposure on the coast near the Orange-San Diego county line at San Onofre. The rock is very coarse-grained, with huge, angular blocks up to 5 feet or more across. It was deposited in the sea, but the angularity and lack of wear on the big blocks show that it was deposited very close to a steep and rugged shoreline. In the Santa Cruz Island exposure, there are some fossil oysters in growth position on a few of the big blocks, further evidence that the blocks did not move much after they fell or rolled a short distance into the sea. Many of these large blocks look as if they had literally fallen off a steep cliff into water deep enough to have been out of reach of wave action.

One of the more distinctive rock types represented among the big blocks (clasts), is blue glaucophane schist. This rock is fairly common in the

Figure 59. Sierra Blanca, southwestern Santa Cruz Island; southwestly view. The light-colored rocks are Miocene Blanca Formation, composed mostly of tuffaceous (ash) sandstones of volcanic origin.

Franciscan Formation at many places in California, but in the southern part of the state is only exposed on the Palos Verdes Peninsula near San Pedro and on Santa Catalina Island. The breccia, on the other hand, occurs at a number of places including Anacapa, Santa Cruz and Santa Rosa islands. Because none of the exposures of the San Onofre Breccia are very near existing exposures of blue glaucophane schist, geologists think that the breccias were derived from former nearby high ridges now either eroded away, submerged, or perhaps covered by younger deposits.

The Blanca Formation, about the same age as the Monterey Formation and the volcanic rocks at Tranquillon Mountain on the mainland, is a good deal less enigmatic than the San Onofre Breccia, but nonetheless quite interesting. It is of note because it is full of glassy, sand-sized, pumice grains indicative of nearby explosive activity. This contrasts with the numerous ash beds found in the Monterey Formation that also indicate the presence of explosive volcanism, but Monterey ash beds are typically very fine-grained suggesting that the air-borne ash had been carried quite a distance from the volcanoes that produced it. It should not be forgotten that during the Miocene, there were many volcanoes in southern California, some of which erupted explosively from time to time, so pinpointing the exact location of the vent that produced the pumice in the Blanca Formation is highly unlikely.

Figure 60. Looking west from East Point along the south coast of Santa Rosa Island. The Miocene Beechers Bay Formation in the foreground is capped by sand in marine terrace deposits of Pleistocene age. Monterey shale of Miocene age is exposed in the cliffs in the middle distance.

Santa Rosa Island

Like Santa Cruz Island, Santa Rosa Island is bisected by a similar east-west fault with several miles of left slip. The relationship of these two island faults is uncertain because the Santa Cruz Island fault does not continue across the intervening channel in a straight line to join the Santa Rosa Island fault. Instead, there is some sort of an offset between these two faults that is concealed beneath the sea.

Unlike Santa Cruz Island, there is no central valley on Santa Rosa and the island generally has lower elevation and less relief than Santa Cruz Island.

The north side of Santa Rosa Island has a broad, well-developed set of marine terraces cut into the middle and lower Miocene marine sedimentary rocks. These terraces are covered in many places by young dune sand. South of the fault, the same sort of rock is present, but in addition there are some exposures of the reddish, non-marine Sespe Formation of Oligocene age as well as exposures of marine Eocene rocks probably partly equivalent to the Cozy Dell Formation in the Santa Ynez Range on the mainland (Figure 60). On the island these rocks extend from Ford Point on the east almost to Sandy Point on the west.

A patch of volcanic rocks occurs between Black and Soledad mountains. These volcanic rocks are mostly basaltic volcanic breccias and include **sills** and **dikes**. Some of the sills and dikes cut the lower part of the marine Monterey Formation which in turn is overlain by the volcanic breccia. This shows that the volcanic rocks are somewhat younger than the lower Monterey because they not only cut across that formation, but some lie on top of it as well.

This is yet another indication of the extensive volcanic activity that affected Santa Barbara County and much of western north America during the middle part of the Miocene. Fluid lavas are represented by the intrusive dikes and sills and the solid eruptive products by ash and other fragmental material that is referred to as volcaniclastic.

Pleistocene dune and marine deposits on Santa Rosa Island have yielded an array of vertebrate fossils, the most famous of which are the bones of pygmy mammoths (*Mammuthus imperator exilis*) whose occurrence on several of the islands has resulted in much controversy. For many years it was assumed that the mammoths could have reached the islands only via a dry land connection, but more recent studies show rather conclusively that during the lowest sea levels of the Pleistocene, mammoths very probably swam across the narrow strait near present Point Hueneme in Ventura County to a single large island known as Santarosae, which then incorporated the four channel islands of today.

Had a land bridge between this large island and the mainland existed at any time during the Pleistocene, most of the 127 or so mainland vertebrates species would have colonized the islands and most of them would probably still be present. Certainly a few of these animals would have disappeared as rising sea levels reduced the size of their habitat and cut them off from the mainland. Further, the arrival of early man, perhaps 20,000 years ago, would have eliminated a few others. However, only 12 or 13 species are present today, far too few to sustain the notion of a land bridge. Surely some snakes and rabbits would have managed to survive had they ever inhabited the islands.

Just when early man arrived is uncertain. Most archeologists put the date somewhere between 10 and 20,000 years ago. A presumed hearth site with what appear to be worked stone tools and burned mammoth bones on Santa Rosa Island has yielded a radiocarbon date of more than 40,000 years. But because these are not associated human remains and because no well-authenticated date of human presence this old had so far been found anywhere in the Americas, the date is highly suspect. The oldest dated human remains from Santa Rosa Island are about 12 to 13,000 years old.

Studies of the fossil or paleomagnetic record of rocks on the islands as well as current direction indicators in sedimentary rocks has provided the astonishing evidence that the present island block has rotated into its present east-west alignment from an earlier northwest-southeast orientation much closer to the southern California coast of Orange and San Diego counties. As mentioned earlier, some of the distinctive Eocene pebble conglomerates on Santa Cruz, Santa Rosa, San Miguel and San Nicolas islands seem to be derived from parent rocks in northwestern Sonora in Mexico and delivered to their depositional sites by west-flowing rivers. Supporting such a history are matching rocks of the Poway Formation near San Diego. Further, the curious occurrence of Torrey Pines (*Pinus torreyana*) only on the San Diego County coast near Del Mar and on Santa Rosa Island is consistent with other evidence indicating the Channel Islands once lay adjacent to the San Diego coast.

San Miguel Island

Larger than Santa Barbara Island, but smallest and westernmost of the Santa Barbara County's three Channel Islands, is windswept San Miguel. For the most part, this island is covered by modern or Pleistocene dune sand or by elevated marine terrace deposits and beach rock. Nearly all the exposures of consolidated rock occur along the shore or in the deeper canyons.

Among the more notable rocks are the Miocene volcanic rocks that make up Harris and Bay points to the east, Prince Island in Cuyler Harbor, Castle Rock off the northwest coast and probably some of the other submerged rocky reefs and pinnacle-like Richardson Rock about 6 miles northwest of Point Bennett on the western shore. These volcanic rocks are very like those on the islands to the east. A good many show pillow structure so we know they were erupted as submarine lava flows. The rock at Harris Point is a plug-like structure of siliceous composition called a dacite.

A volcanic plug is a mass of hardened lava that congealed in the vent or throat of the volcano. These plugs often remain long after the less coherent material of the surrounding cone is eroded away. Devil's Tower in Wyoming is an especially fine example of a volcanic plug. The prominent conical hills that extend from the city of San Luis Obispo to Morro Rock in San Luis Obispo County to the north are good examples of dacite plugs. There is also considerable dacite exposed in the western Santa Monica Mountains of Ventura County.

Sedimentary rocks that make up the island's cliffed shoreline are marine deposits of late Cretaceous and Tertiary age and are very similar in age and rock type to examples in the western Santa Ynez Mountains across the channel to the north, as well as to rocks on the other islands to the east.

Sand dunes are especially prominent on this island owing to its extreme windiness and to the presence of a long sandy beach at Simonton Cove on the northern coast (Figure 26). Prevailing winds sweep sand from the beach and drive it inland across the island in streaks and streams with a pronounced northwest-southeast orientation. The northern part of the shoreline at Cuyler Harbor is covered with dune sand that has cascaded down from the upland to the beach. The amount of actual drifting sand on the island varies from time to time depending upon such things as rainfall fluctuation and the presence or absence of grazing animals. Photos taken in 1928 show that sheep had nearly stripped the vegetation from the island. Removal of these animals by the National Park Service has produced a dramatic recovery of the vegetation and a reduction in the extent of active dunes.

Of special interest in the dunes are the so-called "ghost trees," which are calcified tube-like structures a foot or more high that project above the dune sand (Figure 27). These represent fragile, limy casts that formed around plant roots and stems and even entire tree trunks that once lived in the area. In some places, these ghost trees are associated with high density deposits of dead native land snails that inhabited the area when it was covered by vegetation. A very similar occurrence is present on the dune-covered, windswept western end of San Nicolas Island in Ventura County. Probably these features tell us that vegetation was destroyed either by overgrazing or protracted drought, and once the vegetation died or was eaten to the roots, the persistent northwesterly winds began to strip away the exposed sand, leaving behind the more resistant root casts and numerous land snail shells. Oddly, both the fragile root casts and the delicate snail shells have survived what would seem to be severe sand-blasting.

CHAPTER 6

MINERAL RESOURCES

Diatomite

Probably few people think of Santa Barbara County as the site of important mining activity and perhaps many would be surprised to learn that the world's largest diatomite mines (or quarries) are located in the White Hills south of Lompoc (Figure 52). Diatomite is a sedimentary rock composed largely of the remains of single-celled green plants called diatoms. Although the Santa Barbara County deposit is of marine origin, many diatomite deposits elsewhere are fresh-water lake deposits. Some diatoms even live in the soil. The Lompoc deposit consists of a thick pile of nearly pure diatomite beds, collectively more than 1000 feet thick, part of the Miocene Sisquoc Formation. The Sisquoc Formation is rich in diatoms at many other localities, but all commercial mining is near Lompoc.

There are about 5500 species of diatoms so far recognized and most form a "shell" or more properly, a **frustule**, of opaline silica, often in the form of beautiful intricate meshworks (Figure 61). The shapes of diatom frustules vary from circular to quite elongate and usually have two parts like a pill box. When these organisms reproduce by cell-division, each new individual takes half of the frustule and constructs a new half inside the one they inherited. Periodically, diatoms discard both halves of the frustule and reproduce sexually, forming new, normal-sized frustules.

Though tiny, inconspicuous, and not well-known to the layman, diatoms are of enormous importance. They form the basis of the food chain in the sea—the phytoplankton—and are sometimes called "the grass of the sea". Like other green plants, they use carbon dioxide and water to form carbohydrates and give off free oxygen during the process. It has been suggested that half the life on earth is made up of diatoms, and if true, they must also be a major source of atmospheric oxygen. In addition, many diatoms contain a globule of oil which, after burial and much alteration, is likely to be the main source of petroleum and natural gas. There is an almost universal association of organically-rich sedimentary rocks like the Monterey and Sisquoc formations with oil and gas deposits.

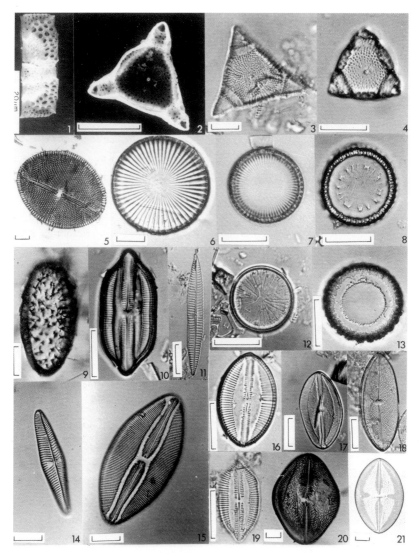

Figure 61. Diatoms representative of the Sisquoc Formation found at the Celite Quarry near Lompoc. The scale bar on each photo is 20 microns or about 0.008 inch. (Photo by J. A. Barron).

As noted previously, diatom abundance in the sea seems to be limited by the availability of iron and silica. Dramatic increases in diatom numbers (blooms) often follow volcanic activity which supplies these two nutrients. Both the Sisquoc and Monterey formations are diatomaceous and contain many thin beds of volcanic ash.

The Lompoc diatomite deposits, the largest in the world, were first mined in 1893 and the present operator succeeded earlier producers in 1928. Diatomite has hundreds of uses and it is likely that virtually everyone in first world countries uses some diatomite nearly every day, or at least uses a product produced with the aid of diatomite.

Individual diatom frustules are so porous that they have enormously large surface areas in proportion to their size. In addition, diatoms do not swell when wet and can absorb up to 75% of their weight in water. The principal uses of diatomite are for various kinds of filtration and clarification, accounting for about two thirds of all uses. Beer and wine are filtered through diatomite. Most swimming pool filters are diatomite-based. The same properties are used in producing cooking oils, sugar and lubricating oil. Diatomite is the usual abrasive in toothpaste and has many other uses as a filler, absorbent material or insulation. Diatomite, in fact, is one of the most efficient heat insulators known.

Modern sea floor diatom oozes offer clues as to the nature of the marine environment when the diatoms in the Sisquoc and Monterey formations were deposited. Oozes are soft, unconsolidated, organically-rich sediments accumulating at a number of places on the modern sea floor. These oozes are usually cool-water deposits, and the largest modern examples occur around the Antarctic continent and at some places in the north polar regions. Smaller deposits occur in the Gulf of California. All these are regions in which upwelling of iron and silica-rich water occurs. For diatomaceous oozes to form, the volume of land-derived muds and sands must be minimal so as not to dilute the organic material with inorganic sediment.

During the time when the Monterey and Sisquoc formations were deposited, volcanic activity in our region was much greater than it is today, and large amounts of silica and iron was introduced into the sea as volcanic ash. Because the nearly pure diatomites near Lompoc cover a small area, the site must have been one in which few streams introduced mud and sand. The site was not polar, so we can be confident that it was the site of intense upwelling.

South of Los Alamos, in the Purisima Hills is the inactive Airox Mine. The rocks there are tar-impregnated diatomaceous shales of the Sisquoc Formation, some of which in the past caught fire naturally and have been baked or even melted. Mining and processing of this exposure of the Sisquoc Formation involves duplicating the natural burning process by roasting or calcining the tarry rock to a temperature of about 1800°F, just short of melting. This produces a bubbly, highly absorbent material sometimes called "expanded shale". Its main uses are as an aggregate for making light-weight concrete and for manufacturing pozzolans, special cements that will harden under water. Other uses of the product include roofing granules, insulation and kitty litter. This mine has been inactive for the past 20 years or so because, among other things, it created air pollution problems.

Metal Mining

Metal mining is not at present important in Santa Barbara County, but it has had an interesting history and the first California discovery of mercury (quicksilver) was made in the upper Santa Ynez River area in 1796.

A small group of mercury mines has operated off and on over the years on the west slope of Cachuma Mountain near Cachuma Saddle. These mines have been variously known as the Eagle Mine, the Acachuma Mine and more recently as the Red Rock Mine. Several tunnels up to 2000 feet long have been driven into the ore body associated with a band of serpentine. Main production occurred between 1874-77, 1916-18, 1935 and in 1938-39. Mining the red cinnabar ore has been intermittent since, chiefly during periods when the price of mercury climbed high enough to warrant more mining. Another similar mining area is along a belt of serpentine about 3 miles long, adjacent to the Little Pine fault not far from Gibraltar Reservoir.

The Santa Barbara County occurrences of mercury are low-grade and have never been as important as the much larger deposits at New Idria in San Benito County nor the New Almaden occurrence in Santa Clara County, though they were discovered before either of these.

Not only are mercury deposits associated with serpentines, but so are small pockets of the principal chromium ore, chromite. No commercial deposits of this mineral have ever been found in the county, though there has been some very minor "recreational" mining in the Figueroa Mountain area.

One does not usually think of Santa Barbara County as gold country, and it is certainly true that there have never been any gold mines of even minor consequence. However, in the late 19th century, the beaches from Point Arguello northward to the Santa Maria River mouth yielded black sands containing gold and platinum. Black sands are placer concentrates of heavy minerals. Most of these sands are composed of iron minerals like magnetite and ilmenite, the latter containing titanium. It is usual for these heavy mineral concentrates to include other heavy minerals as well, and certainly gold and platinum are heavy. The Santa Barbara County beach placers, particularly between Surf and Point Sal yielded $41,000 in gold and platinum in 1889, the peak production year (gold was then selling at about $20 an ounce.)

East of Santa Maria in upper La Brea Canyon in the Sierra Madre Mountains is an old barite mine, the White Elephant Mine. This operation, long inactive, in 1929 and 1930 produced 4000 tons of the heavy mineral barite (barium sulfate) from Cretaceous sedimentary rocks. Barite is used to add weight to oil well drilling muds, to make high density concrete, in glass manufacture and for radioactive shielding.

A little over a mile northeast of Zaca Lake, near Wildhorse Mountain, is a small deposit of copper and zinc associated with a dark intrusive rock that is part of the Coast Range ophiolite. Ophiolite sequences, it will be recalled, are of oceanic floor origin and often associated with subduction zones. Such rocks were more fully described in connection with the Point Sal headland.

South of Cuyama in the northeastern part of the county is the Cuyama Mine. When active during the years between 1965 and 1970, some 18,000 tons of phosphate rock were produced from the Miocene Santa Margarita Formation in Newsome Canyon. This mining operation was discontinued because of a combination of zoning, transportation and dust problems.

The Cuyama phosphate deposit appears very similar to the Pine Mountain deposit in Ventura County, also in the Santa Margarita Formation, where a phosphatic mudstone composed of phosphate pebbles, broken fragments of phosphatic materials such as teeth, bones, shells and interstitial phosphatic cementing material.

Somewhat similar materials are now accumulating on some offshore submarine banks off the southern California coast where upwelling is active. Phosphate deposits are important sources of material for fertilizer and some animal feed supplements.

Among the non-metallic mineral resources produced in Santa Barbara County are rock, sand and dimension stone. Sand and gravel are recovered from the Santa Ynez River near Solvang and from the Sisquoc River near Garey. Sand also has been mined from the eastern edge of the Guadalupe Dunes west of the city of Guadalupe. Dimension stone has been produced at quarries in the Monterey Formation in Tepusquet and Colson canyons east of Santa Maria. The Tepusquet Canyon Quarry was the source of a fine fossil bird displayed at the Santa Barbara Museum of Natural History. The bird, although sometimes referred to as a toothed bird because of serrations on its beak, actually lacks true teeth – as do all post-Mesozoic birds so far discovered. The scientific name of this bird is *Osteodontornis*, meaning bony-toothed bird.

When the towers of the Santa Barbara Mission were rebuilt in 1952-53 because of the poor quality concrete used in reconstructing damage after the 1925 earthquake, stone from Tepusquet Canyon quarries was used.

Although not so far exploited, a very pure, light gray, thick and massive lens of limestone occurs in the lowest part of the Monterey Formation at Bee Rock, just east of Sweetwater Creek and about half way between the range crest of Santa Ynez Peak and Cachuma Lake County Park. The limestone wedge does not occur elsewhere in the Monterey Formation and its origin is puzzling. It is not fossiliferous and appears to be a chemical precipitate resulting from some peculiar, localized submarine process, perhaps one known as a "cold seep" where calcium carbonate-rich water percolates into the sea at the edge of the continental shelf, eventually depositing large masses of calcite or limestone.

OIL AND NATURAL GAS

The Chumash and their predecessors used tar (also called asphaltum or pitch) probably from the time man first arrived in the region between 10 to 20 thousand years ago. It was used as a water-proofing agent for everything from baskets to canoes and probably for many other uses as well, because tar oil seeps were (and are) common, particularly along the south coast and in the channel.

An easily accessible example of an active seep can be seen oozing out of up-ended Monterey shale on the beach just east of the Carpinteria State Park boundary (Figure 62). Numerous examples of inactive seeps occur at other places along the south coast. At More Mesa Beach, east of Goleta Beach Park, are tar-filled cracks in the sea cliff and a tar-cemented rock reef just offshore, favored as a hauling out spot by sea lions when the tide is low.

Active submarine seeps in the Santa Barbara Channel are responsible for the sometimes abundant tar blobs on south coast beaches that have long been the bane of beach users and from time to time have been blamed either on offshore drilling operations or on oil tankers. This tar is a natural consequence of seeps and should not be blamed either on tankering or on offshore petroleum

Figure 62. Carpinteria State Beach Park. The toe of an active tar seep on the beach.

production. In 1776, long before there were any oil tankers or local drilling operations, Padre Pedro Font, while near Goleta, wrote:

> "…much tar which the sea throws up is found on the shores, sticking to stones and dry, little balls of tar are also found. Perhaps there are springs of it which flow out into the sea, because yesterday, on the way, the odor of it was perceptible, and today the scent was as strong as that perceived in a ship or in a store of tarred tackle and ropes."

Similarly, the English explorer, George Vancouver, when he visited the area in 1792 reported :

> "The surface of the sea, which was perfectly smooth and tranquil, was covered with a thick, slimy substance, which when separated or disturbed by a little agitation, became very luminous. Whilst a light breeze, which came principally from the shore, brought with it a strong smell of tar, or some resinous substance. The next morning the sea had the appearance of dissolved tar floating on its surface, which covered the sea in all directions within the limits of our view."

Figure 63. Chinese Harbor on the north coast of Santa Cruz Island. The landslide in the Monterey Formation is caused by a smoldering tar fire which fractures the rock and produces wisps of smoke that are seen near the top of the photograph.

Burning tar, coal or peat deposits are often exceedingly difficult to extinguish, once ignited. They may continue to smolder, sometimes for more than a hundred years. They can be ignited by lightning, brush fires or perhaps accidentally, as has been the case with the peat deposit at Barka Slough near the Vandenberg Air Force Base in the western part of the county. At this writing the peat continues to smolder, giving off prodigious amounts of acrid smoke and so far frustrating ongoing efforts to put it out.

A particularly interesting smoldering tar deposit occurs in the Monterey Formation just above the beach in Chinese Harbor on Santa Cruz Island (Figure 63). This deposit has been smoldering for a number of years and is especially obvious in the winter when warm, moist air rising from the seep helps to create a very perceptible mist or smoke. It is easily located because the heat from the burning has weakened, baked and fractured the rocks, producing a tongue-shaped landslide of rubble that extends to the beach. Tarry material in the fractures in the Monterey Formation provides the fuel, though it is likely that some oil and gas are seeping out of the rock as well.

This burning may bake the associated rock to a brick-red color, or even in some cases fuse or melt the rock to produce a light-weight bubbly black rock that looks superficially like black basalt scoria (a volcanic rock). Old oil and gas seeps that have burned for long periods have produced such baked or fused rocks in many places in the county. For example, at the former Airox Mine south of Los Alamos in the Purisima Hills. Most of these occurrences are located either in the Monterey or Sisquoc formations.

In 1963 the Sandpiper Golf Course was being constructed west of Ellwood, atop the coastal bluffs in an area where there was a smoldering tar seep. In an effort to extinguish the smoldering a bulldozer was used to create a small dam across the mouth of a little canyon so that sea water could be pumped into the basin. When the bulldozer exposed an actively smoldering area, the operator suddenly found himself and his tractor surrounded by flames when abundant oxygen in the air fanned the smouldering fire. Fortunately, flooding by seawater eventually extinguished the smoldering.

In the late 18th and early 19th centuries shortly after the settlement of Santa Barbara, Spanish and Mexican residents reported taking visitors to see the local "volcano". It was a shallow pit from which steam and smoke issued and where the surrounding rocks were baked to a reddish color. It was located about half a mile southeast of Rincon Point in what is now Ventura County, about 50-75 feet above the present railroad tracks. The site is easily visible today and is also marked by an electrically-controlled warning fence between the tracks and the cliff base. The railroad extinguished this burning seep or "solfatara", though local ranchers occasionally still smell sulfurous fumes and it may be that some smoldering persists at depth.

In 1857, a San Francisco druggist attempted to distill coal oil (kerosene) from the Carpinteria tar seep. More recently, Union Oil Company experimented with recovering oil from the petroleum-impregnated Careaga sandstone near Sisquoc. Although they estimated that the Careaga in that area contained 50 million barrels of oil, the process proved uneconomic at the then prevailing crude oil prices.

On the UCSB campus near the Faculty Club is a small, shallow pit surrounded by a low fence. This is all that remains of a former asphalt mine. Mining began about 1890 by the Alcatraz Asphaltum Company. The operators had no background in mining such material and soon found it to be very difficult. At the surface the asphaltum was hard and brittle, but it was soft and very sticky at depth. It oozed through the supporting timbers requiring that mining be continuous so that the shafts would not fill up between shifts. During its heyday, it employed about 50 men and the tar was widely used in California, indeed some was used to pave the streets of San Francisco.

Figure 64. Summerland Oil Field about 1915 showing pioneering offshore production methods. Loon Point is in the middle distance and Rincon Mountain is on the sky line. (Photo courtesy of the Santa Barbara Historical Society).

The mine eventually reached a depth of 550 feet, where petroliferous fumes made the work quite unpleasant, if not actually hazardous. The mine shut down in 1895, not because it was exhausted, but because active surface flows in La Brea Canyon north of the Sisquoc River and east of the hamlet of Sisquoc were cheaper and easier to recover. These La Brea Canyon flows exude from the Monterey Formation.

Nearly all the very early oil fields in Santa Barbara County, and elsewhere for that matter, were drilled near active oil and gas seeps, but by the first decade of the 20th century, geological interpretation began to be more influential. It was soon learned that many oil fields were located on dome-shaped folds in the rocks called anticlines, and many fields were soon located by professional geologists searching for these structural features. Oil drilling, exploration and wild-catting eventually demonstrated the existence of many other sorts of traps.

The largest oil fields in the Santa Barbara area are the modern offshore fields, but if one considers only onshore production, the two most important fields have been Elwood west of Goleta, and the Santa Maria Valley field in the north county. Although Santa Barbara County has been a significant oil and gas producer since just before the turn of the century, its total production

Figure 65. A gusher at Hartnell No. 1, Orcutt Oil Field in 1904. Drilled by Union Oil the well flowed at a rate of 12,000 barrels of oil per day for about 3 months but continued to flow at a reduced rate for another 2 years. The well, popularly known as "Old Maud", eventually produced about 3 million barrels of oil. (Photo courtesy of the California Oil Museum, Santa Paula).

is far behind that of Kern, Ventura and Los Angeles counties. Nevertheless, there are many interesting and unusual features of Santa Barbara County oil fields that are worth mentioning.

First among the twenty or so onshore and offshore oil fields in the county area was the Summerland field where some production began as early as 1894 (Figure 64). By 1896 some wells were operating from spindly little piers offshore making this small field the site of the world's first offshore production. Wells were very shallow and most production came from about 900 feet below the surface from Pleistocene sands. Production inevitably declined and most of the wells were shut down and the piers removed by the 1940's.

Much drilling activity took place in the north county during the period from 1900 to 1910. During this time, fields at Lompoc, Cat Canyon, Casmalia, and Orcutt were all discovered. Most of these were small and produced a thick, tarry oil. Much of the production from Casmalia, for example, was used for road paving material. The Orcutt field, south of Santa Maria, was the site of California's first gusher (Figure 65). Gushers rarely occur today, but in the early days of oil production (The Orcutt discovery was in 1901) drillers often encountered higher than expected pressures when they drilled into the oil reservoir and they lacked modern techniques for dealing with such high pressure. At the Orcutt discovery well, Hartnell No. 1, dubbed "Old Maud", the gusher flowed at a rate of about 12,000 barrels a day for several months. During that early period, the Orcutt field was also the site of the world's deepest oil well which reached a depth of 5000 feet.

The West Cat Canyon Oil Field in the Solomon Hills north of Los Alamos was discovered in 1908. The discovery well began with modest production. However, after being cleaned out by the drillers, the well began to flow at a rate of 6000 to 10,000 barrels a day. The high rate of flow persisted such that after 2 years, it was still flowing at a rate of 1500 barrels a day, a remarkably long period for a free-flowing well.

Another burst of discoveries took place in the late 1920's and 1930's. One of these fields was the little Santa Barbara Mesa Oil Field (Figure 19), discovered just west of the harbor in 1929. This minor field was pretty well exhausted by 1940 except for two or three wells on the north side of Cliff Drive. The last one continued pumping until 1971.

The two largest onshore fields were both discovered during this period. The Elwood field west of Goleta came into production in 1928 and by 1930 was a major California producer, accounting in 1930, for 6% of all the oil production in California (Figure 66). It was depleted by 1980. In 1934, the

Figure 66. Elwood Oil Field, west of Coal Oil Point in December 1936. The sea cliffs are Monterey Formation. The grassy foothills are on the Miocene Rincon Formation and the brush-covered Santa Ynez Mountains are on Vaqueros Formation and older rocks. The crest of the range is mostly on the Eocene Coldwater and Matilija sandstones. (Photo by Spence Air Photos).

other large field, the Santa Maria Valley Oil Field, came into production. It was unusual because much of the oil was not contained in an anticlinal fold, but in what is known as a "stratigraphic trap". In such fields, the oil is usually confined to the upper ends of tilted, permeable beds whose surfaces had been eroded off and later buried beneath impermeable cap rocks. Such concealed traps require sophisticated geologic techniques to locate because they have little or no surface expression. The Santa Maria Valley field was among the earliest to produce from a stratigraphic trap and it remains an important producer.

In 1948, the South Cuyama Oil Field in the far northeastern corner of the county was discovered. In that same year, modern offshore production began with an oil well located just off Coal Oil Point near Goleta. This was the first California production from a nearshore oil well. The first California production from an offshore platform was in 1958 at Summerland. Since then, about a

Figure 67. Platform A in the Santa Barbara Channel. Aerial view shortly after the blowout began on 29 January 1969. The main leak occurred around the platform itself, but after about an hour a secondary leak developed east of the platform. This leak eventually spread to the west of the platform as well. (Photo by T. Putz).

dozen offshore fields have been discovered off the Santa Barbara County coast. Others have been located offshore of Ventura, Los Angeles and Orange counties to the southeast.

The infamous oil spill of 28 January 1969 occurred when drillers lost control of the fifth well being developed from Platform A. This platform produces from the so called "Giant" Dos Cuadras field, unique among giant fields because the top of the anticline containing the oil and gas lay only 300 feet below the sea bottom. A thousand to several thousand feet typically cover other giant fields. The blowout occurred while drilling and when the borehole was largely uncemented. When the drill reached a depth of 3300 feet, the driller lost control of the drilling mud and a blinding oil mist began to pour out of the drill pipe with a deafening roar. The blowout preventer was shut off, but within minutes abundant gas bubbles began to rise from the sea near the drilling platform. The release of oil and gas about an hour later, caused the sea east of the platform to began to "boil" (Figure 67). Leakage expanded during the first 24 hours to a zone east and west of the platform about 1000 feet long. Evidently oil from the deep, high-pressure zones leaked past the drill pipe upward into the low-pressure rocks just below the sea floor, from which it leaked into the sea through many openings. The leaking borehole was finally cemented on 8 February. Something on the order of 12,000 barrels of oil were discharged into the sea during that time, although some estimates were as high as 60,000 barrels. As the oil reached the shore, there was a substantial loss of sea birds and marine invertebrates and coastal rocks and sandy beaches were coated with heavy tarry oil. Although this oil spill was quite serious, it was much less voluminous than a number of others associated with tanker accidents in various parts of the world. Because the blowout occurred in an area in which environmental concerns were particularly strong, it provided a boost to public environmental awareness world-wide.

Many were surprised, however, at the rapidity with which natural processes mitigated the short-term effects of the spill. Within 6 to 8 months, it was difficult to find any evidence of the spill. Birds and other animals were well on their way to full recovery and beach processes, including movement of sand, had largely cleaned up the beaches. Only the rock areas continued to show much evidence of the spill. No doubt a great deal of this rapid recovery was a consequence of the natural discharge of oil and gas from sea floor seeps which has affected the area for millennia. Specialized bacteria have the capacity to break down crude oil and proved to be remarkably efficient in doing so. Furthermore, crude oil slowly oxidizes and gradually forms hardened, largely inert material which is incorporated into the sediments, though it sometimes coats coastal rocks with a sort of natural macadam or tarmac. Dire predictions of long-term damage turned out to be false, though the spill did produce a colossal mess at its outset.

Natural gas seeps are also quite common in the Santa Barbara Channel and along the south coast of both Santa Barbara and Ventura counties. These, incidentally, are also a major source of air pollution. At sea, seeps are easily seen because they cause bubbles when the gas reaches the sea surface. In some places off the coast, bubbling is so evident that patches of the sea of some acres extent look like bubbling soda pop. On land, of course, because the gas has little or no odor, these seeps are much less evident unless they happen to be ignited, as has happened on many occasions in the past. Most oil seeps also yield natural gas. An offshore seep a short distance off Coal Oil Point is a particularly active one, and after an oil company installed a submarine "tent" to capture the oil and gas as part of a pollution mitigation agreement, it was somewhat surprised to discover that about 500,000 cubic feet of natural gas was trapped each day. This and other submarine seeps in the channel release a total of about 5 million cubic feet of gas and about 150 barrels of oil into the sea and atmosphere every day. The oil released amounts to about 55,000 barrels yearly, an amount nearly equal to the high estimate for the 1969 Platform A spill.

There are 3 major oil seep areas off the south coast of Santa Barbara County. One is near Point Conception, while the second is found near Coal Oil and Goleta points. The third and largest of the 3 is more diffuse than the others and extends from Santa Barbara city eastward to Rincon Point. Collectively, these seeps are estimated to release 50-70 barrels of oil daily into the channel waters. Production from these seeps has been observed to increase dramatically during strong earthquakes, when the entire channel may be covered with an oil slick.

As noted above, these gas seeps also pollute the atmosphere and account for almost half of the air pollutants in the Santa Barbara area. Nearly all of the rest of the local atmospheric pollution comes from the operation of motor vehicles. Anything that can be done to capture or reduce the natural releases will contribute to the reduction of air pollution. Interestingly, offshore oil and gas production lowers reservoir pressures and thereby reduces gas and oil pollution of the sea and air.

Active gas seeps had long been known east of the mouth of Goleta Slough and as a result of exploratory drilling in the area during 1929, the La Goleta Gas Field was discovered. Gas in this small field was depleted by the late 1930's. During the war years demand for natural gas in southern California greatly expanded as a result of defense industry and electric power generation needs. Pacific Lighting Gas, the principal supplier, was in need of a large storage facility to accommodate peak demands, so in 1942, it began to use the depleted La Goleta Gas Field as underground storage. The volume of gas that could be held in this underground reservoir equaled a surface facility about 400 feet high and about 18 city blocks square. This is the equivalent of 160 million cubic feet of gas at a pressure of 2000 pounds to the square inch. Although the top of the La Goleta gas reservoir is about 4000 feet below the surface, occasional leaks have occurred, manifested by bubbles in the water of Goleta Slough. Such leaks have been quickly sealed, not only for safety reasons, but for economic and public relations considerations as well.

Recently, evidence has been found that indicates the presence of very large amounts of what are called methane clathrates or gas hydrates buried in the sediments of the Santa Barbara Channel. These deposits are unstable combinations of methane gas and water, kept in a solid state by a combination of high pressure and low temperature. Changes in either pressure or temperature can cause huge, sudden releases of methane gas, something that may have occurred several times during the Pleistocene Epoch. Because methane is an important greenhouse gas, abrupt, large-scale releases of this gas into the atmosphere may well have caused rapid climatic change. Further, systematic and controlled use of these clathrates, may at some future time, provide an important source of energy.

CHAPTER 7

FEATURES OF SPECIAL INTEREST

Tecolote Tunnel

In response to a prolonged and severe drought in the late 1940s, an effort was mounted by south coast leaders to build an additional, much larger reservoir on the Santa Ynez River to augment water already impounded by Juncal Dam for Montecito and by Gibraltar Dam for the City of Santa Barbara. In order to bring this water to the south coast, a 7-mile long tunnel was bored through the Santa Ynez Range and now delivers water from Lake Cachuma impounded by Bradbury Dam to Glen Annie Reservoir on the south side of the mountains.

Boring this tunnel proved to be extraordinarily difficult. The first contractor was unable to finish the job, so another firm with more experience driving mine tunnels took over and completed the task. Part of the problem was that large volumes of hot water were encountered. These filled the tunnel with a hot mist and made it an astonishingly enervating place to work. The writer made a trip into the tunnel in 1953 before it was completed. At that time something on the order of a million gallons a day of sulfate-rich water was draining out of the tunnel, the air temperature at the working face was 113°F and the humidity 100% despite huge volumes of outside air pumped continuously into the tunnel. Tunnel workers were forced to rest every 20 minutes or so by getting into mine cars full of cool water which provided some respite from the exceedingly oppressive conditions.

Because the maximum thickness of rock above the tunnel is about 2000 feet, these temperatures are not particularly unusual and are roughly consistent with the local rate of temperature increase with depth (the geothermal gradient).

As might be expected, with the huge column of water draining out of the tunnel, springs that had persisted for many years on the surface of the mountain in the vicinity of the tunnel quickly went dry as their source of water was drained from below. For some months, water draining out of the tunnel was allowed to flow into the sea because there were no facilities for storing it. In drought-conscious Santa Barbara, this was very disturbing to many people who had just lived through one of the more severe, but recurring dry periods.

Nojoqui Falls

About 6.5 miles south of Solvang, just off Alisal Road, is a county park featuring a pretty little waterfall about 100 feet high. The falls plunge over resistant sandstone of the Cretaceous Jalama Formation. Spray from the mineralized water has deposited, and continues to build a shield of travertine on the rocks, reminiscent of some deposits of similar composition that occur in limestone caves. The falls occur on upper Nojoqui Creek, as tributary of the Santa Ynez River, and are fed mainly by a spring emerging from the overlying Anita Formation, augmented in winter by runoff from rain.

Montecito Overturn

West of Santa Barbara the rocks on the south face of the Santa Ynez Range are inclined (dip) southward toward the sea as a result of the homoclinal structure of the mountains. Eastward, folding is more intense and near Gibraltar Road some of the rocks are vertical (Figure 8). North of Montecito, much of the rock strata on the south face of the mountains is overturned and dips northward into the mountains instead of toward the sea (Figure 5). In this area the older rocks seem to rest on the younger ones, opposite to the order in which they were originally deposited. The overturned structure extends as far east as Rincon Creek. A similar overturned structure occurs north of the Ojai Valley in Ventura county and is known as the Matilija overturn.

Gaviota Gorge

When U.S. Highway 101 leaves the south coast and turns inland toward Buellton, it passes through the narrow, scenic gorge called Gaviota Pass – the only well-defined break in the Santa Ynez Range west of the Ventura River (Figure 12). The Gorge is so narrow that the northbound lanes of the highway use a tunnel driven through the thick, pale tan, late Eocene Gaviota Formation.

Southbound lanes follow Gaviota Creek, a small stream that has managed to cut the gorge entirely through the backbone of the Santa Ynez Range and which drains a sizeable area extending inland about 6 miles from its mouth at the Gaviota State Beach.

From Gaviota Pass, the highway climbs to Nojoqui Summit, about 925 feet above sea level, but no longer marked by a highway sign. This summit is the drainage divide between Gaviota Creek to the south and the Santa Ynez River to the north. At Nojoqui Summit, a recent landslide in the truck parking area east of the highway has exposed the trace of the Santa Ynez fault which divides a short distance eastward into two branches. If one drives east from Nojoqui Summit along the old coast highway, after about half a mile of bends and curves, this old road cuts across the Santa Ynez fault. The fault follows the narrow, grassy valley on the right ,with sycamores along its trace. Crumbly black Cretaceous shale is present on the mountain side of this valley and white-weathering Monterey shale is exposed in the road cut on the other side of the valley.

Before the present four-lane highway was built, a small settlement called Las Cruces was located at the junction of State Highway 1 to Lompoc and U.S. Highway 101. Nearby are some warm springs that at one time were used as a local spa of sorts. The springs still exist and emerge along the south branch of the Santa Ynez fault where it crosses Hot Springs Creek, just a short distance southeast of the site of Las Cruces.

Figueroa Mountain's Black Smoker

In the late 1970s, oceanographers aboard a deep submersible vessel were exploring the crest of the East Pacific Rise, a sea floor spreading center where two great oceanic plates are separating from one another and generating new crustal material between. The general character and major features of these zones had, of course, been known from earlier surveys and sampling, but when these great sutures were seen for the first time at close range, there were many surprises. Among the remarkable sights were features that were named "Black Smokers", chimney-like vents several yards wide and high, out of which poured very hot waters charged with hydrogen sulfide plus black sulfides of copper, zinc, arsenic and other heavy metals. Even more astonishing were the animals clustered around these seemingly lethal areas where the environment is dark, hot, and loaded with the poisonous gas hydrogen sulfide. The animals, many species previously unknown, included large red and white tubeworms, some more than 4 feet in length, large mats of white bacteria, and various bivalves and crabs, all dependent, apparently, on these hot, presumably poisonous discharges.

Just a few years ago, a local geologist, Robert Gray, examined an old small copper deposit on the southwest side of Figueroa Mountain. He recognized that it was almost certainly a fossil black smoker (Figure 43), formed during Jurassic or Cretaceous time when Franciscan Formation rocks were being deposited on the sea floor at or very near a spreading center. Close examination of the area subsequently revealed the presence of fossil tubeworms and other organisms very similar to the living black smoker communities. Sadly, this remarkable feature, perhaps the only one of its kind yet discovered in North America, has been largely destroyed by vandals. However, some broken fragments have been rescued for further study.

Guadalupe Dunes

The large tract of coastal sand dunes mentioned earlier extends from near the mouth of the Santa Maria River inland about 10 miles (Figure 24). The larger, more extensive part of this same dune tract lies north of the Santa Maria River in San Luis Obispo County. The portion in Santa Barbara County has most of the common dune features (apart from numerous dune-dammed lakes) and has been somewhat less disturbed by human activities.

The dunes may be divided into three more or less distinct ages, as mentioned earlier. The older dunes, generally farthest inshore and appreciably stained by iron minerals, are known as the Orcutt sand (Figure 25). The next younger dunes are light tan, and the youngest dunes, generally not far from the beach, are nearly white and often, particularly north of Mussel Rock, are essentially continuous with the beach from which they are derived (Figure 24).

The dune sand probably averages about 25 feet thick and is seldom more than 100 feet thick. Much of the sand blankets elevated marine terraces and near Mussel Rock, the youngest sand resting on a high terrace has a maximum elevation of about 450 feel above sea level.

The entire tract of dunes, both north and south of the county line, shows very clearly the effects of strong northwesterly winds that have, since Pleistocene time, driven the sand inshore from the beach. Tongues and ridges with an obvious northwest-southwest trend are very evident, as are parabolic dunes, hairpin-shaped features whose downwind ends are ridges of sand formed by "blow-outs" where dune vegetation is sparser or has been disturbed so that bare sand is exposed to the raking effect of the wind.

Growth of Mission Ridge, Santa Barbara

Edward Keller and his students at UCSB recently have been tracing former courses of Mission Creek after it leaves its canyon near the Old Mission and flows across the coastal plain toward the sea. They have discovered that

Mission Ridge is continuing to rise slowly along the Mission Ridge fault and is gradually extending itself westward, causing Mission Creek to be deflected bit by bit in that direction in order to get around the end of the rising barrier (Figure 14). This is yet another reminder that our portion of California and the Transverse Range Province continue to be squeezed by north-south crustal shortening.

Geologic Control of Vegetation

Rocks and soils from which they are derived of course influence the vegetation that grows on them. Sometimes minerals in the rocks are toxic to certain kinds of plants, resulting in distinctive plant communities made of species tolerant of the particular toxic mineral. Serpentine soils are rich in magnesium and many plants are intolerant of such soils. However, Blue Oaks (*Quercus douglasii*), Sargent Cypress (*Cupressus sargentii*), Gray or Digger Pine (*Pinus sabiniana*), Santa Barbara Jewelflower (*Caulanthus amplixicaulus* var *barbarae*) and Needlegrass (*Achnatherium* sp.) are characteristic of plant communities that do tolerate serpentine soils in the county. A good example of such a community is present on the southwest side of Figueroa Mountain, especially on the University of California Sedgwick Ranch Reserve (Figure 55).

Shallow-rooted annual grasses are more tolerant of poorly drained clay-rich soils than are perennial woody shrubs. As a result, one of the more striking geologic controls of vegetation can be seen on the seaward side of the Santa Ynez Range west from Goleta to Point Conception.

Rock units in this area dip seaward and form nearly continuous bands along the range front. The lower foothills are underlain near the coast by the Monterey Formation, on which clayey, adobe-like soils develop. These soils are filled with numerous chips of white, siliceous rock from the bedrock below. Upslope is the Rincon Formation, a mudstone that weathers to a heavy, black clay soil that shrinks and cracks when dry and becomes nearly as hard as concrete. When wet, it expands and becomes am incredibly sticky mud. Both these rock units are normally grass-covered, but underlying them and uphill from them are the Vaqueros sandstone and Sespe Formation, both yielding sandy or silty, rapidly draining soils. The Rincon-Vaqueros contact is sharply evident along this coast because the grassland ends abruptly at the contact and the sandier soils uphill are thickly covered with woody chaparral. There are few areas anywhere where geological control of vegetation is so obvious. Fine examples can be seen on either side of U.S. Highway 101 as it turns inland toward Gaviota Pass (Figure 12).

Sea Cliff at Isla Vista

The most dramatic place to observe the effects of shoreline erosion in Santa Barbara County is at Isla Vista where there has been intense urban development. Various protective structures like sea walls, drains, concrete piers and so on, provide an excellent opportunity to observe that progressive retreat of a sea cliff (Figures 68-70).

Although direct wave attack plays some role in cliff erosion here, it is not the dominant cause, as is easily demonstrated by the continuing rapid retreat of the cliff face behind "protective" sea walls.

Highest tides and storm waves do permit waves occasionally to attack the cliff base and help remove any accumulated loose debris, but most of the time nonmarine processes dominate. These include rainwash on the face of the cliff, the effects of burrowing animals, foot traffic up and down the cliff, growth of vegetation on the cliff, water draining over the cliff face and salt weathering.

Salt weathering occurs when sea spray is driven against the cliff by winds, soaks into the porous rocks there, evaporates and allows tiny salt crystals to grow, wedging apart the rock grains. The process is slow, inconspicuous, but persistent and produces a constant sloughing of fine material.

The rate at which cliffs retreat varies considerably and some of the differences observed are difficult to explain, at least in the short term. As a rule, the higher the cliff, the faster it will retreat, other things being equal. In the Santa Barbara-Goleta area, the highest rates of cliff retreat are observed at the 80-85 foot high cliffs at More Mesa (Figure 53) where the annual retreat rate over the past 50 years has been about a foot. For the coast between Hope Ranch and Santa Barbara Harbor, the average rate is less, about 7-8 inches annually. At Isla Vista, near the UCSB campus, the rate is closer to 6 inches. However, during the very wet and stormy winter of 1982-83 some cliffs in the Santa Barbara area retreated as much as 18 inches in a single storm. Similar effects were observed after the 1998 El Niño storms.

Some of the erosional processes can be managed to a degree, but over the long term, erosion will inevitably undermine and ultimately destroy any buildings near the cliff top. Shorelines are among the most dynamic of any natural environment.

Figure 68. Isla Vista in February 1987. Note the erosion of the sea cliff behind the sea wall.

Figure 69. Isla Vista in February 1993. Continuing erosion of the sea cliff behind the sea wall; note the extensive exposure of the supporting columns.

Figure 70. Isla Vista in April 1997. Further erosion of sea cliffs behind the sea wall; note especially the extent of erosion under the houses on the right.

ROAD LOGS

In following the road logs, remember that odometers on vehicles will vary somewhat, so your mileage may not always agree precisely with what is given in the logs. Make appropriate corrections from time to time.

The main highways now have emergency telephone call boxes at regular intervals. These signs may be helpful in checking your mileage. At the bottom of each Call Box sign are two numbers. The one on the left refers to the highway number, and the one on the right to the mileage from where the highway starts, or enters the county. On Figure 71, the numbers tell you are on U.S. Highway 101, 16.7 miles from the Ventura County line at Rincon Creek. Unfortunately, perhaps, the signs on the other side of the highway use the same point of origin, not the San Luis Obispo County line near Santa Maria, so they are less useful when travelling in the opposite direction.

In addition, all state and federal highways have paddle signs along the shoulder (Figure 72). The top number in the illustration refers to State Highway 154. The letters below are the county designation, and the large numbers below show miles and hundredths from where the highway begins or enters the county. In the illustration, State Highway 154 begins at its junction with U.S. Highway 101 near Los Olivos and the sign is located 28.90 miles from that point.

Remember too, that on freeways, offramps at a given junction may be a half mile or more apart, depending on direction of travel. The road logs use the actual crossing, even if it is on a bridge or below one.

Finally, keep in mind that some features along the road are much more easily seen from one direction of travel. So, if you happen to be going in the opposite direction, the feature may be harder to see.

U.S. Highway 101 from Rincon Creek to the Santa Maria River

For the first 45 miles, this highway is seldom out of sight of the sea, but after turning north through the scenic gorge of Gaviota Creek, the highway remains well inland from the coast until it reaches Pismo Beach in San Luis Obispo County to the north. The entire route, however, is very attractive. The south coast affords many fine views of the sea on the one hand

Figure 71. Call box sign on U. S. Highway 101 in Santa Barbara. The sign shows the distance from the Ventura-Santa Barbara county line.

and the steep wall of the Santa Ynez Mountains on the other. North of Gaviota there are still stretches of the road that pass through ranching country with few settlements, though vineyards are increasingly occupying former grazing lands along the highway.

0.0 (91.2) Enter Santa Barbara County from the south at Rincon Creek. The road climbs up a small hill from near sea level at Rincon Point to the surface of an elevated Pleistocene marine terrace about 150 feet above sea level.

0.2 (91.0) Rocks in the road cut to the right (east) are parts of the Sisquoc and Monterey formations, both showing a north dip. They are overturned here.

0.7 (90.5) Where the highway swings to the west, near the top of the hill, the Carpinteria fault is crossed. On the seaward side of the road, atop the terrace, there was formerly a small depression,

Figure 72. Paddle sign on State Higiway 154. 28.90 miles from its junction with U.S. Highway 101 near Los Olivos.

          used at one time as an auto racetrack (the Carpinteria Thunderbowl). This depression is a sag pond or pull-apart structure on the fault (Figures 28 and 54).

2.1 (89.1) Looking straight ahead at the lower part of the Santa Ynez Mountains, a low ridge can be seen in front of the main range. Between the high crest and the lower ridge is the trace of the Arroyo Parida fault, one part of a string of faults, or trend, that extends from Ventura County westward to the UCSB campus.

3.7 (87.5) Cross Santa Monica Creek. This is one of the streams that drains into Carpinteria Salt Marsh to the left (south).

4.3 (86.9) Good view of Carpinteria Salt Marsh. The salt marsh is one of the larger, relatively undisturbed coastal marshes in Southern California. Part of it is managed by the University of California's Natural Land System as a teaching and research reserve. Other parts are managed by the city of Carpinteria and the two homeowners associations, Sandyland Cove and Sandyland.

Like all salt marshes, geologically speaking, this is a very temporary feature. In the natural course of events, sediment carried from the mountains into the marsh by Santa Monica and Franklin creeks will eventually fill it and convert it to dry land, barring a rise of sea level or some such tectonic event.

4.7 (86.5) At the coast, more or less behind some of the commercial buildings in the village of Santa Claus, can be seen the towers and domes of a white house with distinctive Moorish architecture. This was the site of some of the most severe beach erosion that occurred in the late 1930's as a result of sand impoundment at Santa Barbara Harbor to the west. In one particularly bad year, a width of 180 feet of beach was cut away, damaging part of this unique building.

5.0 (86.2) The Santa Ynez Mountains on the right (north) rise steeply here along the Arroyo Parida fault, which, as before, lies behind the first ridge adjacent to the Carpinteria plain. The higher parts of the mountains show rocks dipping into the mountain face, the Montecito overturn. In the lower foothills, planted with avocados, can be seen the red soils derived from the Sespe Formation. In the overturn area, older rocks rest on younger rocks, demonstrating intense folding.

5.5 (85.7) Arroyo Parida bridge.

7.9 (83.3) Just beyond Toro Canyon, the headland to the left (south) is Loon Point. The beach beyond, also to the left (south) is the site of the old Summerland Oil Field (Figure 64). The rocks making up Ortega Hill, straight ahead, belong to the Casitas Formation. Both Loon Point (Figure 30) and Ortega Hill are the result of anticlinal folding. The old Summerland Oil Field between these two points produced oil from this anticlinal fold.

8.9 (82.3)	Just beyond the crest of Ortega Hill, scenic Fernald Point comes into view on the left (south). This small headland is the mouth of Romero Creek and it is of deltaic origin.
11.8 (79.4)	Salinas Street offramp. Just to the left (south) out of sight, is the Andree Clark Refuge, all that remains of a once extensive marshy area. Much of the industrial part of Santa Barbara's lower East Side is on filled ground reclaimed from the marsh.
12.7 (78.5)	The residential area in the city on the lower foothills is known as "The Riviera" or "Mission Ridge", and behind this ridge toward the higher Santa Ynez Mountains beyond is a depression eroded along the Mission Ridge fault, the probably western extension of the Arroyo Parida fault mentioned previously.
	The main part of the city lies in a topographic depression between Mission Ridge and the Mesa Hills to the southwest.
13.2 (78.0)	The Mesa Hills with high points at TV Hill and Lavigia Hill are straight ahead at this point.
13.85 (77.4)	Cross State Street underpass. Just west of here is the buried Mesa fault, concealed by stream alluvium.
16.1 (75.1)	The highway rises slightly onto a patch of bouldary old alluvium deposited in late Pleistocene time. The Earl Warren Show Grounds on the right (northeast) are on this deposit as are the low hills to the left (southwest). The higher hills to the south near La Cumbre Middle School are on the Santa Barbara Formation.
16.7 (74.5)	The Arroyo Burro water gap (Figures 9 and 13). On the left, an obvious canyon cuts across the Mesa Hills to the sea beyond. The canyon, currently occupied by the rather small Arroyo Burro Creek, very likely was earlier traversed by a larger stream such as Mission Creek.
17.4 (73.8)	The Sespe Formation is exposed in the low hill on the left (south) near the La Cumbre Road overpass. The buried junction of the Mesa, Mission Ridge and More Ranch faults is near here. The hills of Hope Ranch on the left (south) are bounded by the More Ranch fault, along which they have been raised. The northern part of these hills is underlain by the Santa Barbara Formation.

On the seaward side of Hope Ranch (not visible from the highway) are marine terraces cut into the Monterey Formation. To the west, the most prominent of these terraces has an elevation of 80-100 feet; in Hope Ranch this same bench is nearly 200 feet above sea level in some places (Figure 53).

18.9 (72.3) Gravelly older alluvium rests on Santa Barbara Formation on both sides of the highway here, though the Santa Barbara Formation is more easily seen in the road cuts on the access road along the highway on the right (north).

20.9 (70.3) Cross Maria Ygnacio Creek, one of the seven streams that drains into Goleta Slough which lies to the left (south) between the Airport and the elevated terrace on which the UCSB campus is situated.

Goleta Slough, as it exists today, is a small remnant of what was formally an extensive salt marsh, reaching almost to the junction of Hollister and Fairview avenues. During World War II, the Santa Barbara Airport was built by the military largely by excavating material from Mescalitan Island and from the seacliffs near Ellwood. Mescalitan Island is a small hill near the mouth of the slough and is the site of the Goleta Sanitary District's sewage treatment and water reclamation plant. The "island" is an erosional remnant left by streams cutting channels on either side of it, San Jose and Atascadero creeks to the east and San Pedro, Los Carneros, and Glen Annie creeks to the west. It was earlier noted that the confluence of these streams and their tributaries during the Pleistocene glacially-lowered sea level, cut a bedrock channel several miles off the mouth of the present slough, which has since been completely filled with beach and nearshore deposits (Figure 32).

24.1 (67.1) At the Glen Annie Canyon offramp, the highway rises onto a patch of older alluvium that is evident in the low hill on the right (north).

27.9 (63.3) Low hills to the right (north) are made up of Monterey Formation.

30.0 (61.2) Cross Dos Pueblos Creek. The hills on the right (north) are mostly composed of the mudstone from the Rincon Formation.

33.1 (58.1)	A nice example of badland erosion can be seen in the railway cut on the left (south) in the older alluvium.
33.7 (57.5)	Northbound exit to El Capitan State Beach. The mouth of Cañada del Capitan Creek is a small boulder delta (Figure 34).
34.6 (56.6)	Remnants of the former El Capitan Oil Field on the right (north). For many years there was a single wooden oil derrick on the beach pumping from this field. Many of the creeks for the next few miles have fairly narrow valleys where they cross the Monterey Formation to the right (north) of the highway. Farther upstream, where the creeks cross the less resistant, underlying Rincon mudstone, the valleys widen noticeably. This is particularly evident in Refugio and Tajiguas creeks to the west. These broader areas are often planted with lemons or avocados. Still farther upstream, where the creeks cross the Vaqueros Formation sandstone and conglomerate, the valleys narrow sharply and the grassy hillsides are replaced by thick woody chaparral (Figure 50).
	The seaward side of the Santa Ynez Mountains west of Santa Barbara is made up of a sequence of sedimentary rocks dipping seaward, forming very marked east-west bands. The harder sandstones are prominent ridge-formers and provide numerous exposures of bare rock easily seen from the highway. The softer shales, mudstones and siltstones have fewer outcrops and form smoother, less craggy surfaces.
	Although the sequence of rocks on this side of the Santa Ynez Mountains is usually in normal order with the oldest rocks near the crest, there are a number of faults and folds that trend more or less parallel to the mountains, but are not evident to the traveler along the highway. The higher ridges between San Marcos and Gaviota passes are mostly Eocene sandstones of the Matilija and Sacate formations, but at the Gaviota Pass Tunnel, the prominent sandstones belong to the late Eocene Gaviota Formation (Figure 12).
36.6 (54.6)	Cross Refugio Creek.
38.2 (53.0)	The sea cliffs that can be seen ahead are formed from Monterey Formation.

38.9 (52.3)	The bare rock exposed in the mountains ahead are mostly Gaviota Formation sandstones, the grass-covered foothills are developed on Rincon Formation (nearest the mountains) and the Monterey Formation closer to the shoreline. The boundary between these two units is difficult to see from a distance.
39.8 (51.4)	Exposure of Monterey Formation cherty shale in the road cut.
41.1 (50.1)	Cross deep valley at Arroyo Hondo Bridge.
41.7 (49.5)	Note the sharp boundary between the grass-covered Rincon Formation and the chaparral covered sandstones of the Vaqueros and Sespe formations to the right (north).
43.6 (47.6)	The steeply inclined sandstones near the mountain crest belong to the Gaviota Formation, a marine deposit of late Eocene age.
44.7 (46.5)	Mariposa Reina offramp at the oil processing plant.
45.5 (45.7)	The village of Gaviota. It is situated on a sloping, uplifted marine terrace about 150 feet above sea level.
47.0 (44.2)	Gaviota Pass Tunnel passes through spectacular exposures of the Gaviota Formation (Figure 12).
47.5 (43.7)	Cross south branch of the Santa Ynez fault at about where the highway begins to swing right (east). About half a mile to the northwest, also on the fault is Gaviota or Las Cruces Hot Springs, reached by a foot trail in Gaviota State Park. This was formerly a traveler's rest stop.
48.4 (42.8)	Intersection of State Highway 1. This locality is close to the site of the former small settlement of Las Cruces, obliterated when the highway was converted into a freeway.
50.5 (40.7)	Cross Gaviota Creek. Rocks on the left (west) belong to the Monterey Formation, those on the right (east) to the Gaviota Formation.
50.9 (40.3)	Nojoqui Summit. The north branch of the Santa Ynez fault is crossed here. It is exposed in the roadcut to the right (east) at the truck parking area. The Rincon Formation is north of the fault and the Vaqueros Formation to the south.
52.9 (38.3)	Note the small landslide in the Rincon Formation on the left (west).

Road Logs

54.7 (36.5) Good exposures of the Cretaceous marine Espada shale. Although the north slope of the Santa Ynez Mountains as far as the Santa Ynez River Bridge is complicated by several east-west trending faults and folds, the ages of the rocks get older as the river is approached. The Espada Formation is one of the older rocks exposed in the county.

56.6 (34.6) The Santa Ynez River Bridge. The river valley here is almost 1.5 miles wide, but narrows both upstream and downstream. The three dams on the river all lie up stream from this point, so the river bed is often dry during the summer unless water is being released from Cachuma Lake. Beyond Buellton, the highway follows the course of Zaca Creek to the crest of the low, anticlinal Purisima Hills and thereafter descends into the synclinal valley of San Antonio Creek.

The bedrock of the Purisima Hills is mostly Monterey Formation, capped here and there with a soft, light-tan to whitish marine rock of Pliocene age called the Careaga Formation.

62.6 (28.6) Junction with State Highway 154 to San Marcos Pass. Rocks on either side of the highway here belong to the nonmarine Plio-Pleistocene Paso Robles Formation. For about the next 5 miles, the highway follows the axis of the Los Alamos syncline and another unnamed syncline beyond Alisos Road. A careful observer can see rocks dipping gently toward the valley on either side of the highway. On either side of this syncline are two anticlinal folds.

The anticline on the right (northeast) is the site of the Cat Canyon Oil Field and the one on the southwest (left), the site of the Barham Ranch and Orcutt oil fields.

70.7 (20.5) The town of Los Alamos.

72.6 (18.6) Cat Canyon Road junction. Between this point and the junction with Palmer Road, the highway crosses several exposures of Orcutt sand, a Pleistocene dune deposit.

78.5 (12.7) At 10 o'clock on the left (west), is Mount Solomon, an oddly-shaped low mountain formed on a resistant limy zone in the Careaga sand and capped with a small patch of nonmarine Paso Robles Formation. It is easily recognized by the plethora of telecommunications towers on its summit.

80.3 (10.9)	Orcutt sand is exposed on both sides of the highway here, and underlies the irregular brush-covered hills nearby.
81.8 (9.4)	The highway is now on the older dune sand, a younger deposit than the Orcutt and probably mostly of Pleistocene age. Where undisturbed, these old dune sands are mostly grass-covered.
85.7 (5.5)	The road here leaves the dune sand and passes onto river alluvium deposited by the Sisquoc River. These are some of the premium farm lands of the Santa Maria Valley.
90.9 (0.3)	Junction with State Highway 166 to Guadalupe.
91.2 (0.0)	San Luis Obispo County line. The county line was laid out when the course of the Santa Maria River lay south of its present position. The River Bridge is entirely in San Luis Obispo County.

Refugio Pass Road from U.S. Highway 101 to State Highway 246 at Solvang

Although paved on the south face of the Santa Ynez Mountains, this road is winding and steep. It makes several crossings of Refugio Creek, some in culverts, but others on the road surface. The road down the north side of the mountains is unpaved, steep and very rough as far as Quilota Creek. It is generally impassable in wet weather, and in recent years has been closed by the County Road Department.

0.0 (14.8)	The offramp on U.S. Highway 101. Some Rincon mudstone is exposed near the offramp, but inland the road passes onto the familiar Monterey Formation.
0.1 (14.7)	Note the big slump on the right (east) developed in the Monterey Formation.
0.4 (14.4)	Cross Refugio Creek. The Monterey Formation is exposed in the roadcut on the right (east).
0.5 (14.3)	The valley of Refugio Creek, widens here as the road passes onto the softer, more easily eroded Rincon Formation. As noted, this pattern is repeated in many of the stream valleys draining the western part of the Santa Ynez Range. It is especially evident in the valley to the west, Tajiguas Creek.

1.9 (12.9)	Cross the Refugio fault. The fault lies just north of a small east-west ridge of Vaqueros sandstone that crosses the valley. The fault repeats the rock sequence so that on the uphill side (north) of the fault, a narrow band of Rincon Formation is present.
2.3 (12.5)	Prominent east-west ridge of Vaqueros sandstone. The canyon narrows sharply at this point. Good exposures of the Vaqueros on either side of the road.
2.9 (11.9)	Sespe Formation sandstone exposed here.
3.3 (11.5)	Gaviota Formation sandstone exposed on the left (west). The Anita shale is present between the Gaviota and the Sespe formations, but is not well exposed here.
3.4 (11.4)	Entrance to Circle Bar-B Guest Ranch on the right (north-east). Near this point the road crosses onto the Sacate Formation, and remains on this formation all the way to the crest of the mountains. Exposures are not particularly good anywhere along the road.
7.2 (7.6)	Summit and beginning of unpaved portion of the road down the north-facing side of the mountains. This slope is heavily forested and exposures of the various rock units are quite poor. The rocks at the summit belong to the Sacate Formation of Eocene age. As one continues down hill, older and older units are crossed, first the Cozy Dell shale, then the Matilija sandstones followed by the Anita shale, a little more Matilija and where the road reaches the Quilota Creek crossing and the pavement begins, the main trace of the Santa Ynez fault is crossed. Quilota Creek makes a sharp bend toward the right (east) at the fault.
10.5 (4.3)	Begin paved road.
10.6 (4.2)	Cross the Santa Ynez fault. The high, grassy hill on the right (east) is underlain by Rincon mudstone.
11.1 (3.7)	The high hill on the right (east) with oaks on the skyline is underlain by the Monterey Formation. Where the canyon narrows, the road crosses a branch of the Santa Ynez fault.
11.3 (3.5)	A good exposure of the Rincon mudstone can be seen in the steep stream-cut cliff to the right (east). Such good exposures of the non-resistant rock are not common.

13.0 (1.8) Santa Ynez River Bridge.

14.4 (0.4) Exposure of river gravels in an older, high river terrace.

14.8 (0.0) Junction of Refugio Road and State Highway 246.

West Camino Cielo Road from San Marcos Pass Road (State Highway 154) to Refugio Pass Road

This road provides a number of spectacular views of the coastal plain to the south, the south face of the Santa Ynez Mountains, the islands across the Santa Barbara Channel, and north to the Santa Ynez River Valley and the San Rafael Mountains beyond. The first 4 miles of the road are paved and easily traveled, but the next 9 miles are progressively rougher, particularly near Broadcast Peak. The road is not for the faint-hearted, nor for low-clearance passenger cars, though it can be negotiated by ordinary passenger cars with care and if one is not in a hurry.

0.0 (18.8) Junction of Kinevan Road and San Marcos Pass Road (State Highway 154). The rocks in this area are nearly flat-lying Coldwater sandstones that form the crest of the broad Brush Peak anticline.

0.1 (18.7) Prominent rocks exposed along the road belong to the marine Coldwater Formation of Eocene age.

0.2 (18.6) Junction of Kinevan Road and West Camino Cielo Road. Take the left fork.

1.0 (17.8) Boulder-pile weathering can be seen in massive Coldwater sandstone. The rectangular blocks that are formed by planes of stratification and jointing at right angles, are gradually rounded as weathering attacks the corners from three sides, the edges from two, and the faces from just one.

1.5 (17.3) Stacked boulders are the result of weathering of thick beds of Coldwater sandstone, which are nearly horizontal here. Fine examples on the left (south) of the road.

2.1 (16.7) More examples of boulder-pile weathering of thick Coldwater sandstone. Such huge blocks have, in the past, been swept down many canyons by mudflows during incredibly intense storms, accounting for the huge boulders such as those in Mission Canyon at the Botanic Garden. In this area, boulders would be carried down San Pedro Canyon instead.

3.5 (15.3)	Prominent stack or tower of Coldwater sandstone can be seen on the right (north). Good examples of Coldwater sandstone also exposed in the road cuts on the left (south).
3.8 (15.0)	Junction with the road to the Winchester Gun Club. Good examples of Coldwater sandstone on the left (south).
4.1 (14.7)	Begin unpaved portion of the road. This point affords good views of the Santa Ynez River Valley and the higher county of the San Rafael Mountains to the north.
4.2 (14.6)	Leave the Coldwater sandstone and pass onto the Sacate Formation. The Sacate is composed of shales and thin-bedded sandstones.
4.3 (14.5)	The high peak in the distance with radio and telecommunication towers is Broadcast Peak (4028 ft). The high peak just beyond is Santa Ynez Peak (4298 ft), also with towers. Santa Ynez Peak is the highest point in the range west of Carpinteria.
4.4 (14.4)	Just beyond the switchback in the road, the route passes onto the Cozy Dell Formation, a dark-colored marine shale with some thin sandstone beds. The Cozy Dell weathers to fine chips and is characterized by numerous talus piles at the base of cliffs or road cuts. It, like the Coldwater and Sacate, is of late Eocene age. Because it weathers to fine-grained material, unpaved roads crossing this formation are notably smoother than those crossing resistant sandstones.
4.5 (14.3)	The high ridge above the road to the left (south) is capped with resistant Coldwater sandstone resting on softer Sacate and Cozy Dell shales.
4.6 (14.2)	Cozy Dell shale with thin sandstone beds exposed on the left (south).
4.8 (14.0)	The fine-grained talus piles along the road are from Cozy Dell shales exposed by the road cuts.
5.3 (13.5)	Intersection with minor side roads to right and left. Keep straight ahead.
5.8 (13.0)	Deep road cut. Spheroidal weathering can be seen in the Cozy Dell shales. Similar examples are seen on East Camino Cielo Road.

6.2 (12.6)	Deep road cut. Cozy Dell Formation is exposed on both sides of the road. Just beyond this cut are good views of the high country to the north as well as the coastal plain to the south.
7.5 (11.3)	Cobble conglomerate lens in the Cozy Dell Formation. The larger light-colored rocks are granitic. Nearly all the cobbles are well-rounded, indicating considerable transport from distant sources as well as a nearshore depositional site. Coarse materials such as this are very unlikely to be deposited beyond the surf zone. The cobble source may be as far away as the Mojave Desert, or even northwestern Mexico.
8.5 (10.3)	Cross onto the middle Eocene marine Matilija Formation. Because much of this formation is a massive sandstone, the quality of the road surface deteriorates for the next 5 miles.
9.6 (10.3)	Fine views both to the north and south.
11.4 (7.4)	The road on the right leads to the telecommunication towers atop Broadcast Peak.
12.9 (5.9)	Santa Ynez Peak on the right (north) above the road, though the summit is not visible from this point.
13.3	Begin paved road. The branch road to the right leads to the summit of Santa Ynez Peak.
16.3 (2.5)	The Matilija Formation seen here is mostly shale with thin sandstone beds. It is exposed on the right (north).
16.5 (2.3)	Cross a narrow band of Cozy Dell shale.
17.5 (1.3)	Sacate Formation shale exposed in the road cut on the right (north).
17.8 (1.0)	Sacate sandstone beds here.
18.4 (0.4)	Sacate shales with thin sandstone beds.
18.8 (0.0)	Junction with Refugio Pass Road.

State Highway 1 from U.S. Highway 101 at Las Cruces to Guadalupe

This road is one of the more scenic routes in Santa Barbara County. It passes through undeveloped ranch lands from U.S. Highway 101 almost to State Highway 246 near Lompoc, providing a pastoral scenary much more common 50 years ago or more, before southern California experienced the post-World War II population explosion.

Road Logs

0.0 (50.8)   Head of northbound offramp on U.S. Highway 101. For the first mile the road crosses several rock units of Eocene age. Near the crest of the hill, the road is on Matilija sandstone. None of these rock units is particularly well displayed here.

1.2 (49.6)   Cross the trace of the north branch of the Santa Ynez fault. You will see a linear valley to the left (south) of the road that lies approximately along the trace of this major fault. As the highway descends into the valley of upper El Jaro Creek, it follows the contact between the nonmarine Sespe Formation on the right (north) with the marine Alegria on the left (south).

2.7 (48.1)   The white sand in the road cut on the left (south) is Alegria Formation sandstone.

4.5 (46.3)   Cross El Jaro Creek. The high ridge on the left (south) is formed on the resistant Matilija sandstone.

5.8 (45.0)   The rocky ridge to the left (south) is, again, Matilija sandstone.

6.8 (44.0)   Cross El Jaro Creek.

7.5 (43.3)   Entrance to Rancho San Julian to the right (east).

8.5 (42.3)   The brush and oak-covered ridge to the left is capped here by the early Miocene Vaqueros Formation. The ridge is developed on a canoe-shaped (synclinal) fold, and the northern two-thirds of the fold is formed from cherty Monterey shales.

8.6 (42.2)   Cross El Jaro Creek.

10.1 (40.7)   Cross El Jaro Creek.

11.0 (39.8)   Road cuts on the left (west) are exposures of marine Espada Formation of Cretaceous age. El Jaro Creek occupies a narrow gorge on the right (east).

11.7 (39.1)   Small landslide on the left (west).

11.9 (38.9)   Fine exposures of cherty Monterey shale in the big road cuts in this area. North of the road, in the overlying Sisquoc diatomite, is the Grefco Quarry, now inactive. This operation is not visible from the highway.

12.1 (38.7)   Cross El Jaro Creek.

12.3 (38.5)   Good exposures of the Monterey Formation in the deep road cuts.

to 12.7 (38.1)	Junction with the Jalama Road. StateHighway 1 now follows
13.7 (37.1)	the drainage of Salsipuedes Creek, of which El Jaro Creek is a major tributary.
14.4 (36.4)	Cross the bridge over Salsipuedes Creek.
14.8 (36.0)	On the left (west) near the top of the ridge is an old, overgrown mine dump from the Celite diatomite quarry operation to the west. This dump was last active in the late 1940's.
15.5 (35.3)	More old mine dumps near the crest of the hills to the left (west). Patches of white Monterey Formation are exposed in the chaparral-covered hills. Between this point and the last one mentioned above, the road crosses the axis of the San Miguelito syncline which downfolds the Sisquoc into the underlying Monterey. Both the Celite Quarry to the west and the Grefco Quarry to the east exploit Sisquoc diatomite from this synclinal fold.
15.9 (34.9)	Cross the axis of the anticlinal fold just north of the San Miguelito syncline. The trend of both these folds is more or less east-west and at right angles to the road.
18.0 (32.8)	Junction with State Highway 246. Turn left (west) toward the town of Lompoc.
19.3 (31.5)	Turn right (north) on H Street, following State Highway 1.
21.3 (29.5)	Santa Ynez River Bridge. Just beyond the bridge, the road rises about 75 feet or so onto an older Santa Ynez River terrace.
21.6 (29.2)	Older river gravels can be seen in the low bluff on the right (east).
22.1 (28.1)	Junction with Harris Grade Road (formerly the route of State Highway 1) to the right (east). Keep left. This junction is close to the contact between the river terrace deposits and the Orcutt sand that covers much of the area to the north. The Orcutt is a Pleistocene dune sand deposit. This gently undulating area crossed by the highway is known as Burton Mesa.
23.8 (27.0)	Road to Vandenberg Village to the right (east).
25.4 (25.4)	Junction with San Lucia Canyon Road to the left (west).
26.6 (24.2)	Prominent patches of white sand on the distant hills to the right (east) are exposures of Orcutt sand.
28.6 (22.2)	Utah Gate, Vandenberg Air Force Base. Turn right (east).

Road Logs

29.9 (20.9) Junction with Firefighter Road. Road cuts in this area show good examples of badland erosion of the soft, weakly cemented Orcutt sand.

30.7 (20.1) Road descends into the valley of San Antonio Creek and leaves the Orcutt sand, under which is poorly-exposed Sisquoc Formation, before reaching the alluvium of the valley floor.

31.2 (19.6) Junction with San Antonio Road from the left (west).

31.7 (19.1) Junction with San Antonio Road East from the right. The low-lying marshy valley floor to the right (east) is a wetland known as Barka Slough.

34.9 (15.9) Junction with State Highway 135 on the right (east). The road now follows stream alluvium deposited by Harris Creek. The base of the hills on either side of the highway is the marine Careaga sand capped with nonmarine Paso Robles Formation, especially on the right (east) side of the valley.

37.1 (13.7) The road crosses the divide between the south-flowing Harris Creek and the north-flowing Orcutt Creek. The canyon to the north is called Graciosa Canyon. At the summit, exposures of the marine Careaga Canyon can be seen on the left (west).

40.6 (10.2) Junction with State Highway 135. Keep left on State Highway 1.

42.2 (8.6) Junction with Black Road. Low road cuts here expose Orcutt sand.

44.5 (6.3) The high ridge on the left (southwest) is the Point Sal Ridge, a part of the Casmalia Hills. The highest peak is Mt. Lospe (1640 ft) which is underlain by the Monterey Formation. The geology of this ridge is of unusual interest. The northern side is composed mainly of Tertiary marine deposits, but with some nonmarine and even a few patches of Cretaceous rocks north of Mount Lospe. The seaward side includes a well-developed, nearly complete ophiolite sequence. This sequence is well exposed along the beach. See the Corralitos log for more details.

47.0 (3.8) Railroad over-crossing. Orcutt sand exposed here.

47.7 (3.1) Junction with Brown Road, the route to Corralitos Canyon.

49.0 (1.8)   The low hills in the distance to the left (west) are Guadalupe sand dunes.

49.5 (1.3)   Enter the City of Guadalupe.

50.8 (0.0)   Santa Maria River crossing.

San Marcos Pass Route (State Highway 154) from U.S. Highway 101 in Santa Barbara to U.S. Highway 101 between Buellton and Los Alamos

This route is one of California's Scenic Highways. As the road climbs the south face of the Santa Ynez Mountains north of Santa Barbara, it provides a number of good exposures of Eocene and Oligocene rocks in the road cuts, though there are few places to stop safely and examine rock features at close range.

The north slope of the Santa Ynez is much more heavily vegetated (except for three or four years after a fire). Because of folding and displacement near the Santa Ynez fault zone, a greater variety of rocks are exposed in north slope road cuts than on the south face.

After the road turns away from the mountains and traverses the Santa Ynez Valley, it crosses young nonmarine deposits, most of which are not well exposed. However, this road provides some good views of the southwest flank of the high San Rafael Mountains.

0.0 (32.4)   Junction of State Highway 154 and U.S. Highway 101 in Santa Barbara. The road begins on young alluvium deposited by Atascadero and Cieneguitas creeks, and just before reaching the junction with Cathedral Oaks-Foothill roads, it passes onto the bouldery fanglomerate of Pleistocene age deposited perhaps 12,000 to 16,000 years ago during an interval of very intense erosion.

0.8 (31.6)   Junction of Cathedral Oaks and Foothill roads.

2.2 (30.2)   The road crosses onto the reddish, nonmarine Sespe Formation chiefly of Oligocene age. Although red is a dominant color in this formation, the upper part includes a number of thick, buff-colored sandstone beds that can be seen in road cuts on the right (south).

3.3 (29.1)   On the right hand side of the road at the apex of a prominent curve in the highway, is an area of Sespe rocks that has been subject to repeated landsliding.

4.0 (28.4)	An exposure of the conglomerate in the lower part of the Sespe Formation can be seen on the right (north).
4.3 (28.1)	A deep road cut exposing the basal, pebbly Sespe conglomerate.
4.5 (27.9)	Sespe-Coldwater contact. The older Coldwater Foundation lies beneath the Sespe Foundation.
5.1 (27.3)	Two small, but prominent faults are exposed in the large road cut on the left (south). The rocks are Coldwater.
5.6 (26.8)	Junction with Painted Cave and Old San Marcos Pass roads.
5.7 (26.7)	Deep, prominent road cut with good exposures of the Coldwater sandstones.
7.2 (25.4)	Junction with Kinevan/West Camino Cielo road on left (west). Note the nearly flat-lying Coldwater sandstone beds in the cliff across the valley to the left (west). These rocks dip very gently toward San Marcos Pass but appear from this vantage point to be horizontal.
8.0 (24.4)	Crest of San Marcos Pass (elevation 2222 ft). Coldwater is exposed in the road cuts. There are several large folds that obliquely cut across the mountains in this area. These account, in part, for the location of the pass itself. The Laurel Canyon syncline lies just northeast of the road, and the more prominent Painted Cave syncline still farther east (see East Camino Cielo Road log).
8.7 (23.7)	Looking across the Santa Ynez River Valley one can see a prominent grass-covered hill (Loma Alta, 2758 ft) in the middle distance. This hill is underlain mainly by the Monterey shale.
9.2 (22.5)	Cold Spring Arch bridge. Coldwater sandstone outcrops can be seen across Los Laureles Canyon to the right (east). The axis of the Laurel Canyon syncline is generally followed by the stream. The sandstones on the far side of the canyon dip toward the valley.
9.9 (22.5)	View Point. If you park where the plaque is set in a sandstone boulder and look north, Loma Alta is in the middle distance. The high ridge behind, with the pine trees on its crest, is Little Pine Mountain (4506 ft), just to the right of Loma Alta. Little Pine Mountain's crest lies in the axis of a large synclinal

fold. The highest point is on Monterey Formation and the rocks that can be seen in the south-facing slope are the Espada shales (Figure 45) of Cretaceous age, as well as sandstones and serpentinites of the still older Franciscan complex. The serpentinites are greenish and easily seen. The far side of the Franciscan rocks lies along the trace of the Camuesa fault, and the near side, at lower elevation, marks the trace of the Little Pine fault. The topography of the band of Franciscan rock is more irregular and hummocky than either the rocks above or below (Figure 55).

The high peak behind Loma Alta and slightly to the left (east) is San Rafael Mountain (6593 ft), the second highest peak in the county (after Big Pine Mountain, 6828 ft).

10.7 (21.7) Beyond the Paradise Road junction, the highway crosses a number of narrow stream valleys separated by ridges made up, for the most part, of Sisquoc and Monterey Formation shales, capped here and there with patches of bouldery older alluvium. Cachuma Lake, the largest of three reservoirs on the Santa Ynez River, can be seen on the right (north).

11.9 (20.5) Cross the trace of the Santa Ynez fault. It is not exposed here but concealed beneath stream terrace deposits.

14.6 (17.8) View of Figueroa Mountain, straight ahead in the distance. Like Little Pine Mountain, Figueroa Mountain's summit lies close to the axis of a synclinal fold in the Monterey Formation, with Cretaceous marine sedimentary rocks underneath. Lower on the hillside below the Cretaceous, is a band of Franciscan rocks, much the same as those on the lower slopes of Little Pine Mountain (see Figueroa Mountain loop log).

15.3 (17.1) Cherty Monterey shale exposed in road cuts on both sides of the road.

16.1 (16.3) The very soft, white, weakly consolidated sandstone seen in the road cut, particularly on the left (south), is the shallow-water, marine Careaga Formation of late Pliocene age, and shows how far inland the sea reached about 2-3 million years ago.

Road Logs

17.8 (14.6)   Entrance to Cachuma Lake County Park.

20.6 (11.8)   Prominent buff-colored cliffs on the north shore of Lake Cachuma are mostly shaly rocks of the Monterey and Sisquoc formations.

22.1 (10.4)   Santa Ynez River bridge. The road here is on Santa Ynez River deposits, which extend beyond to the Santa Agueda Creek bridge. After this point, the road makes a straight run across a flat-lying sheet of Pleistocene older alluvium. In a few of the deeper gullies cutting this surface, there are some exposures of the Monterey Formation.

24.3 (8.1)   Junction with State Highway 246 to Solvang, Buellton and Lompoc. At this junction, if you'll look to the right (north) of straight ahead, you will see a prominent pyramidal peak, Zaca Peak (4341 ft). Like both Little Pine and Figueroa mountains, this peak lies near the axis of a synclinal fold in the Monterey Formation.

26.5 (5.9)   Junction with Baseline Road. The grassy hills on the right (northeast) with scattered oak trees are formed on the nonmarine Paso Robles Formation of Plio-Pleistocene age. The Paso Robles is an old stream and flood plain deposit full of whitish or buff-colored chips of the Monterey Formation.

29.8 (2.6)   Alamo Pintado Creek bridge.

30.5 (1.9)   The hilly area just beyond the junction with Foxen Canyon Road is formed from Paso Robles Formation. If you look closely to the right (north), you can see parts of the old railroad bed for the tracks of the narrow-gauge Pacific Coast Railroad that terminated at Mattei's Tavern in Los Olivos and operated from Port San Luis (Port Harford) from the early 1880's until the 1930's.

32.4 (0.0)   Junction with U.S. Highway 101.

Lompoc to Honda Valley

This short trip takes one south from Lompoc past the entrance of the large Celite Diatomite Quarry, over a steep ridge and down into Honda Valley to a locked gate at the boundary with the Vandenberg Air Force Base. One cannot travel all the way down Honda Valley to the sea.

0.0 (9.3)	Junction of Ocean Avenue and I Street, in Lompoc.
1.0 (8.3)	Note the contorted and complexly folded Monterey Formation in the road cuts on the left (east).
2.1 (7.2)	Entrance road to Celite Diatomite Quarry.
2.6 (6.7)	Good exposures of cherty Monterey shale across the creek.
3.4 (5.9)	Miguelito County Park entrance on right (west).
4.4 (4.9)	Large, ancient landslide mostly in Monterey Formation dominates the hillside on the left side (south) of the road. Note the irregular, somewhat hummocky surface of this large slide that extends nearly 600 feet above the road.
4.6 (4.7)	The grassy slope across the valley on the right (west) is mostly underlain by the Cozy Dell shale and shows several well-developed surficial slumps, especially near the base of the hill.
7.5 (1.8)	Junction with Sudden Road. Gaviota-Sacate Formation on both sides of the road.
7.7 (1.6)	Descend into Honda Valley. Much of this valley lies along the trace of the Honda fault, whose south side has moved up with respect to the north side. The fault also has some left slip. As a consequence, older rocks are exposed on the south side of the fault.
8.7 (0.6)	The lower slopes of the hills on both sides of the road are composed of Cozy Dell shale. The Honda fault here lies north of the valley bottom.
9.3 (0.0)	Locked gate and end of the road. The steeply inclined sandstones that can be seen straight ahead belong to the Eocene marine Matilija Formation.

East Camino Cielo Road from Flores Flat Junction to Mono Forest Service Campground

Like the West Camino Cielo Road, this road provides many excellent and panoramic views of the coastal plain, channel, and islands to the south. It also affords excellent views of the high San Rafael Mountains to the north as well as the upper two reservoirs on the Santa Ynez River. The road descends into the Santa Ynez River valley and crosses the stream. It provides several points of access to the San Rafael Mountain wilderness areas and to Blue Canyon on the north slope of the Santa Ynez Mountains.

The road surface is a good deal better than West Camino Cielo Road, though in recent years the pavement has been allowed to deteriorate and one must dodge a lot of potholes. The graded, unsurfaced road east of Romero Saddle is much better than the unimproved portion of West Camino Cielo Road and is readily negotiated by ordinary passenger vehicles except when high water prevents crossing the Santa Ynez River at the fords.

0.0 (19.5) For the first 4 miles from the start of the log at Flores Flat junction, the rocks exposed along the road belong to the marine Juncal Formation of early to middle Eocene age. Most of the Juncal rocks are shales, but there are some sandstones as well.

2.6 (16.9) Montecito Peak at 2 o'clock, somewhat downhill on the seaward side of the crest. This peak is formed from Matilija sandstone.

3.6 (15.9) Cold Springs Saddle. This is the head for the Cold Spring trail that descends the south face of the Santa Ynez Range. The trail northward to Forbush Flat also starts here. A water tank is next to the road on the right (south).

4.4 (15.1) Late Cretaceous marine Jalama Formation exposed along the road. This is mainly a dark-colored shale, but includes some sandstone beds and conglomerates.

6.0 (13.5) The Montecito overturn on the south face of the mountains is visible here. Sandstone beds are steeply inclined to the north.

6.7 (12.8) Water tank on the right (south). Pavement ends here.

7.1 (12.4) The pebbly Jalama Formation is well exposed in road cuts.

8.0 (11.5) Divide Peak Junction. Gate.

8.2 (11.3) The road now closely follows the south branch of the Santa Ynez fault.

8.4 (11.1) Cross south branch of the Santa Ynez fault and onto the Jurassic-Cretaceous Franciscan Formation, which here is mostly graywacke sandstone. The abrupt change in rocks is evidence of the fault.

10.4 (9.1) Cross main Santa Ynez fault.

10.5 (9.0) Blue Canyon Pass. Fifty yards or so to the left (north) of the road is a fine exposure of Franciscan radiolarian chert (Figure 42) which is gray rather than reddish. Just a few yards beyond is a good exposure of the marine, early Eocene Sierra Blanca limestone (Figure 46). Blue Canyon itself extends westward from here and was eroded along the Santa Ynez fault (a fault-line valley).

11.0 (8.5) Cross the Santa Ynez fault and enter an exposure of the marine, late Jurassic to middle Cretaceous Espada Formation, mainly a dark, greenish-gray shale with minor sandstone beds.

11.9 (7.6) Cross the Santa Ynez River at the Juncal Forest Service camp.

12.3 (7.2) The Eocene marine Juncal Formation is exposed on the hillside to the right (north).

15.0 (4.5) Cross Big Caliente Creek at the Pendola Guard Station.

15.4 (4.1) Juncal Shale exposed in the road cut on the right (north).

16.1 (3.4) P-Bar Forest Service Campground.

16.3 (3.2) Early Miocene marine Temblor Formation, a sandstone, exposed on the right (north).

17.2 (2.3) Crest of hill. Monterey shale exposed on the hill directly ahead.

18.5 (1.0) Gate and River of Life sign.

18.8 (0.7) Grass-covered areas here are developed on the Eocene Juncal shale. A landslide can be seen on the right (north) side of the road.

19.1 (0.4) Junction with the trail to the Mono Adobe.

19.5 (0.0) Mono Forest Service Campground. A good exposure of the Eocene Juncal shale can be seen across the creek.

Gibraltar Road from Sheffield Reservoir to the Junction of East Camino Cielo Road and San Marcos Pass Road (State Highway 154)

This road provides views of some of the same rocks encountered along the heavily traveled San Marcos Pass Road (State Highway 154), but because traffic is light, it affords many safe places to stop and look at the rocks at close range, or to view the scenery.

Because this road crosses the western end of the Montecito overturn, some rocks are overturned and others are nearly vertical, unlike the situation at San Marcos Pass, where all the rocks on the south side of the mountains dip seaward.

0.0 (17.9) Start at junction of State Highway 192 and Mountain Drive in Santa Barbara near Sheffield Reservoir. Take Mountain Drive to left (north).

0.2 (17.2) Junction of Mountain Drive and Gibraltar Road. Take Gibraltar Road to the right (east). The big boulders on both sides of the road are part of an alluvial fan of late Pleistocene age. This deposit is called a fanglomerate and was deposited by torrential floods and mudflows as the glacial climates in North America gave way to the warmer post-glacial time. The boulders are nearly all from the sandstones of the marine Eocene Coldwater and Matilija sandstones exposed in the higher parts of the Santa Ynez Mountains here.

1.0 (16.9) Nonmarine red Sespe Formation chiefly of Oligocene age is exposed in road cuts.

1.2 (16.7) Good example of the basal conglomerate of the Sespe Formation on the left (southwest).

1.3 (16.6) Saddle separating Rattlesnake Canyon on the west from Sycamore Canyon on the east. This saddle marks the location of an ancient shoreline about 45 million years ago when the marine environment represented by the Coldwater Formation was replaced by a land environment represented by the red Sespe Formation. If you look closely at the Coldwater just uphill from the saddle, you will find fossil oysters that tell us that this part of the Coldwater was deposited in an estuary close to shore. The Coldwater here also contains streaks of reddish rocks showing that there was some oscillation between marine and nonmarine conditions as Coldwater time drew to a close. This is not at all unusual in such transitional periods; geologic change seldom moves smoothly and consistently in one direction. In this area, both the Coldwater and Sespe formations dip steeply north, into the mountains, and are overturned. This is the western end of the Montecito overturn, which continues eastward as far as the county line with Ventura County.

2.6 (15.3) Massive Coldwater sandstone exposed in the road cuts here.

2.9 (15.0) Beginning of the Cozy Dell marine shale of Eocene age. Iron gate on the left (west).

3.3 (14.6) After passing the switchback in the road, there is a good exposure of the Cozy Dell shale on the right(east).

4.9 (13.0) Nearly vertical beds here are within the Cozy Dell Formation. Most of this formation is shaly, but it includes a number of sandstone beds as well. This particular exposure is interesting because the beds are nearly vertical and some are covered with shallow-water ripple marks (Figure 8).

5.0 (12.9) The marine, Eocene Matilija Formation begins here. In the hand specimen of this rock are nearly indistinguishable from the Coldwater, but they are easily separated by their position in the rock sequence.

5.4 (12.5) The Juncal shale of Eocene age begins here at Flores Flat. It forms more subdued topography than the more resistant Matilija sandstone, but similar to Matilija it was deposited in the sea. Like the Cozy Dell, the Juncal forms swales and valleys in contrast with the ridge-forming Coldwater and Matilija sandstones. La Cumbre Peak, for example, is on the Matilija Formation (Figure 43).

6.8 (11.1) Junction of Gibraltar Road and East Camino Cielo. Keep left (west) on East Camino Cielo Road.

7.9 (10.0) Nice example of spheroidal weathering in Juncal mudstones on the right (north). This onion-like structure develops as the weathering process attacks blocks of rock divided into more or less rectangular chunks by bedding planes and joint cracks. Corners are attacked from three sides, edges from two, and faces from one, gradually changing the rectangular blocks into rounded ones. It is especially evident where the weathering shells remain in place as they do here. The road remains on the Juncal Formation beyond Angostura Pass, where a gated road descends to the north. At this point, the Santa Barbara city's Mission water tunnel lies about 2000 feet directly underneath the road. Angostura Pass is one of several places along East Camino Cielo Road where one can view the upper Santa Ynez River, Gibraltar Reservoir and the San Rafael Mountains beyond. Greenish rocks can be seen north and northwest of the reservoir. These are exposures of the metamorphic rock serpentinite, which is exposed along the Little Pine fault that passes under the lake. Other exposures occur along the Camuesa fault to the northeast. These serpentinites are composed mainly of the mineral serpentine and are part of the Jurassic-Cretaceous Franciscan Formation that makes up much of the county's basement rock. The high peaks to the north are Big Pine Mountain (6828 ft), the highest point in the county, and Madulce Peak (6541 ft), just to the east. Both these peaks are underlain by Eocene marine rock.

8.0 (9.9) We enter the Matilija Formation here, which makes up much of the higher part of the range in this area.

8.7 (9.2) Side road to La Cumbre Peak (3985 ft) lookout. A short walk to the summit gives fine views of the coastal plain and the city of Santa Barbara and environs as well as the channel and offshore islands. In very clear weather five or six of the eight offshore islands are visible from here. La Cumbre Peak is an exposure of the resistant Matilija sandstone, and is the highest peak near Santa Barbara, though the range crest is higher north of Carpinteria where Divide Peak reaches 4690 feet, and to the west of San Marcos Pass where there are two high peaks, Broadcast Peak (4028 ft), bristling with telecommunication towers, and Santa Ynez Peak, about a mile west at 4298 feet.

10.8 (7.1) Because of the Painted Cave synclinal fold that lies obliquely across the range, the road crosses younger and younger rocks as the core of the fold is approached. We are now in the Cozy Dell Formation, which covers the Matilija below. The younger Sespe occupies the center of the fold. Sespe rocks are present down the north face of the range from the Painted Cave settlement, and also northwesterly almost 3 miles to near the Santa Ynez River. The Coldwater sandstones, of course, lie between the younger Sespe and the older Cozy Dell.

11.8 (6.1) Head of Arroyo Burro trail down the south face of the range.

12.0 (5.9) Exposure of the Coldwater formation. Both the Cozy Dell and Matilija formations are deeply buried beneath the Sespe here.

	The road descends steeply beyond this point on the Sespe Formation.
14.3 (3.6)	Sespe Formation here. Between this point and the Painted Cave Road, the Sespe contains a number of yellow-brown sandstone beds in addition to more reddish rock.
15.8 (2.1)	Painted Cave Road junction.
16.0 (1.9)	The basal conglomerate of the Sespe is exposed in road cuts on the left (south) side of the road. Some pebbles (clasts) are about a foot across and composed of granitic rock. This is an especially interesting exposure because it includes a number of rock types from light-colored granitic rock to dark quartzites and metavolcanic rocks. The granite pebbles and boulders, though they appear sound from a few feet away, have been thoroughly rotted by weathering and one can easily stick a pocket knife into them. Don't try this with the darker metavolcanic and quartzite pebbles as they are, after perhaps 30 million years or so, still hard and fresh. The source or sources of these pebbles pose some interesting questions because there are no granitic, quartzitic or metavolcanic rocks in our part of California. The original source was probably somewhere as far away as the Sierra Nevada, the Mojave Desert, or elsewhere. Possibly a much closer source was later buried and is no longer exposed. Also, there is a slight possibility that some of the more durable pebbles were just recycled from older conglomerates like some of those in nearby Cretaceous units. Field studies, however, suggest recycling is improbable.
16.2 (1.7)	Coldwater Formation here.
17.9 (0.0)	Junction with San Marcos Pass Road.

Jalama Road from State Highway 1 to Jalama Beach County Park

This road goes through attractive ranch and farming country in the southwestern corner of the county to Jalama Beach County Park where one comes out of the canyon of Jalama Creek about in the middle of the large southwest-facing bight between Points Conception and Arguello. Some of the road hugs the trace of the Pacifico fault, the westernmost unit of the Santa Ynez fault system.

0.0 (14.3)	Junction of Jalama Road and State Highway 1.
0.3 (14.0)	Almost immediately upon leaving Highway 1, the Jalama road curls around the nose of a hill made up of Monterey Formation, followed by a small patch of Cozy Dell shale on the south side of the hill just before the descent to the flat alluvial deposits of Salsipuedes Creek and its tributaries. Just beyond, the road crosses a low ridge extending into the valley from the west. This ridge is made up of Cretaceous sandstones and shales and forms the north wall of the valley of La Hoya Creek, which enters Salsipuedes Creek from the right (west).
1.2 (13.1)	On the south side of La Hoya Creek, the road crosses an exposure of steeply dipping marine Cretaceous Espada shale and sandstone that can be seen on the left (east) side of the road.
2.1 (12.2)	Bridge over Salsipuedes Creek. Once the road starts to climb out of the valley toward Jalachichi Summit, it passes over a sequence of marine Cretaceous sandstones and shales of the Jalama Formation. Near the start of the climb, the road crosses the axis of the Pacifico anticline. Rocks on either side of the valley to the north dip northerly, and those near the summit dip to the south.
5.7 (8.6)	Jalachichi Summit. Rocks here belong to the Eocene marine Matilija Formation.
6.0 (8.3)	The brush-covered hills at about 2 o'clock across the valley are developed on the Jalama sandstone; the grassy area at about 1 o'clock is on the more poorly-drained Miocene Rincon mudstone, a relationship between vegetation as well as sandstone and mudstone seen at many places in the county.
7.4 (6.9)	The road passes through a narrow gap that marks the location of the Ramajal syncline. Youngest rocks lie in the core of this fold and are exposed of either side of the road and are Monterey cherty shales. Beyond the gap, toward the main part of the valley of Jalama Creek, rocks become progressively older as one leaves the synclinal axis.
8.5 (5.8)	The rocks on either side of the valley here are Cretaceous and part of the Jalama Formation. The creek here closely follows the trace of the Pacifico fault.

10.1 (4.2)	Side road on the right (north) leads to Jalama Ranch headquarters. The headquarters buildings are on the Rincon mudstone.
12.2 (2.1)	Good exposures of Monterey cherty shale in road cuts until the road reaches the flattish uplifted marine terrace followed by the railroad.
12.6 (1.7)	Steeply inclined Monterey shales with some tarry beds.
12.8 (1.5)	Fine views of the sweep of coastline from near Point Conception on the south to Point Arguello on the north. Most of the seacliffs along this coast are formed from Monterey and Sisquoc formations of Miocene age. Just before the road swings left (south), it crosses the axis of the Jalama anticline. The high peak visible to the northwest at the railroad crossing is Tranquillon Mountain (2170 ft). Tranquillon Mountain is formed from the Tranquillon volcanic rocks that are about the same age as the Miocene Monterey Formation, with which they are closely associated.
14.3 (0.0)	Jalama Beach County Park. There are some small, modern sand dunes here on the seaward side of the park. The bedrock underlying the park area is mainly Sisquoc Formation, and the flat uplands are elevated marine terraces of Pleistocene age.

Santa Ynez, Santa Rita and Lompoc Valley Route
State Highway 246 from State Highway 154 to Surf

This road connects four towns, Santa Ynez, Solvang, Buellton and Lompoc, and roughly parallels the Santa Ynez River. It passes through important agricultural land and ends on the exposed western coast of the county at Surf.

0.0 (35.6)	Junction of State Highways 154 and 245. From this point, the road crosses nearly flat-lying older alluvium of Pleistocene age to near the town of Santa Ynez, where it descends slightly onto the younger alluvium of Zanja de Cota Creek for a short distance, climbing back onto the older alluvium as far as the valley of Alamo Pintado Creek, which, once again, is on modern stream alluvium.

1.8 (33.8)	Chumash Casino on left (south).
2.5 (33.1)	Junction with Refugio Road to Refugio Pass.
4.9 (30.7)	Mission Santa Ines on the right (south).
5.8 (29.8)	Descend onto the younger alluvium of the Santa Ynez River. This alluvium represents a somewhat elevated part of the river flood plain and is slightly older than the alluvium on either side of the present stream channel. It is a stream terrace.
8.3 (27.3)	Junction of U.S. Highway 101 and State Highway 246 at Buellton.
9.0 (26.6)	The flat-toped hills on the right (north), slightly above the stream terrace, are formed from older alluvium of Pleistocene age.
14.4 (21.2)	Intersection with Drum Canyon Road to the right (north). The low hills on either side of the road here are on the Orcutt sand, the oldest of the three dune deposits in western Santa Barbara County. The Orcutt Formation is of Pleistocene age.
16.4 (19.2)	Small lake on the right (north) side of the road.
18.2 (17.4)	Junction with Campbell Road. Between this point and the next junction with the Campbell Road loop, the road rises over a low ridge of Paso Robles Formation. From here westward, the road remains on the younger alluvium of the Santa Rita Valley, deposited by several local creeks.
20.3 (15.3)	The road climbs slightly onto an exposure of Orcutt sand.
20.5 (15.1)	Junction with Tularosa Road. The road is on Orcutt sand here.
24.8 (10.8)	Santa Ynez River Bridge. Hills on the left (east) are composed of Sisquoc Formation at the base, but Older Alluvium at the top.
25.1 (10.5)	Enter the city of Lompoc at the junction with 7th Street.
27.4 (8.2)	Brush-covered hills to the left (south) are chiefly Sisquoc Formation.
30.3 (5.3)	Low hills to the right (north), across the river are mostly Orcutt sand with some Monterey exposed in a few gullies.
34.5 (1.1)	The hills to the left (south) are developed mainly on the Orcutt sand, but with a few exposures near the base of the hills of Sisquoc Formation.

35.1 (0.5) The older dune sand forms the surface between the road and the higher hills to the east that are developed on the Orcutt sand. The older Dune is intermediate in age between the Orcutt sand and the modern dune sands along the coast. The base of the hills on the left (south) have some small exposures of Monterey and Sisquoc formations.

35.6 (0.0) Parking lot at Surf.

Paradise Road from State Highway 154 to locked gate near the Santa Ynez River just beyond the Live Oak Forest Service Picnic Area.

This road follows the Santa Ynez River in an area that lies about equidistant between Cachuma Lake on the west and Gibraltar Reservoir on the east. It is one of the more popular Forest Service recreation areas in the county. The log includes a short side trip to the Upper Oso Forest Service Campground.

0.0 (10.6) Junction of Paradise Road and State Highway 154. The first part of this road more or less follows the boundary between an old Santa Ynez River terrace on the left (north) and exposures of the marine Eocene Cozy Dell shale on the right (south).

0.6 (10.0) Good exposures of older alluvium atop the Cozy Dell shale.

1.1 (9.5) Little Pine Mountain straight ahead, in the distance.

1.4 (9.2) Loma Alta, a high grassy hill, can be seen at about 9 o'clock on the left (north). Loma Alta has an elevation of 2758 feet and is developed on Monterey Formation.

2.5 (8.1) Fremont Forest Service Campground. Just uphill on the right (south) is the Sespe Formation, but it is not easily seen from the road. Between Fremont Camp and the river crossing just beyond Los Prietos Ranger Station. The road crosses various river terrace deposits, all Pleistocene or younger.

4.5 (6.1) Los Prietos Ranger station. The high cliffs across the river are marine Miocene Monterey cherty shales.

5.8 (4.8) Junction with Upper Oso Campground road to the left (north). From this point to the next river crossing the road is on Monterey Formation, and on the north side of the river.

7.5 (3.1) Second river crossing. The Rincon Formation is on the north side of the river, and the older Vaqueros on the south. Both are marine Miocene deposits and both are older than the Monterey Formation. The hillside on the left (north), just beyond the crossing, is Monterey cherty shale.

8.5 (2.1) The high cliffs on the right (south) are Vaqueros sandstone.

9.2 (1.4) Third river crossing at Live Oak Picnic Area. The cliffs on both sides of the river are composed of the marine Miocene Temblor sandstone. This rock, absent on the south coast, is intermediate in age between the Rincon mudstone and the Monterey shale.

9.4 (1.2) Fourth Santa Ynez River crossing. The prominent cliff at 10 o'clock on the left (north) is part of the Monterey Formation as are other exposures of buff-colored rocks nearby. The prominent ridge is Temblor sandstone.

9.9 (0.7) Fifth Santa Ynez River crossing.

10.0 (0.6) Sixth Santa Ynez River crossing.

10.4 (0.2) Live Oak Forest Service Picnic Area. The light-colored cliffs on the left (north) side of the river are Monterey shales. The darker rocks beyond, in the high hill, belong to the Cretaceous marine Espada Formation and rest on poorly-exposed Franciscan serpentinites. The Little Pine fault crosses the river here.

10.6 (0.0) Locked gate.

Side Trip to Upper Oso Forest Service Campground

0.0 (1.1) Junction of Upper Oso Road and Paradise Road. The first half mile of this road passes through a fairly narrow canyon cut in the familiar Monterey cherty shale, exposed on both sides of the canyon.

0.5 (0.6) The high ridge straight ahead is developed on the Eocene Matilija Formation. The same rocks as at La Cumbre Peak in the Santa Ynez Mountains to the south.

0.9 (0.2) Massive sandstone straight head is Matilija Formation.

1.1 (0.0) Forest Service Upper Oso Campground.

Figueroa Mountain Loop

This road provides some of the better views of the Franciscan rocks that form much of the basement of the county. It crosses some of the major faults in the country and penetrates farther into the high San Rafael Mountains than do most other public roads. It also allows access to several old mining areas where copper, chromite and quicksilver (mercury) have been obtained in past years.

0.0 (35.1) Begin at the junction of State Highways 154 and 246. Take the Armour Ranch Road to the right (east). The road first crosses old Santa Ynez River terrace deposits, probably of Pleistocene age. These are referred to as the older alluvium.

1.1 (34.0) Good exposure of the older alluvium in the road cut on the left (north).

1.4 (33.7) The road swings to the right (south), the prominent hill on the right is composed of older alluvium resting on the marine Pliocene Careaga sand on the north side of the hill, and on Sisquoc and Monterey on the south, east and west sides.

1.6 (33.6) Junction with Happy Canyon Road to the left (east). Turn left on this road.

2.0 (33.1) Road descends onto terraces and alluvium of Santa Agueda Creek.

2.4 (32.7) The forested mountain, straight ahead in the distance, is Figueroa Mountain (4528 ft).

3.9 (31.2) After the junction with Alisos Canyon Road, Happy Canyon Road follows along the east side of the stream valley. The higher hills on the right (east), are formed on Paso Robles Formation that contains abundant chips of Monterey shale. It is nonmarine.

7.9 (27.2) If Happy Canyon Creek is dry, as it usually is, note the white coating on the stream cobbles. This results from lime dissolved in the stream water that is left behind as the water on the rocks evaporates. This is a widespread phenomenon in Santa Barbara County in late summer and fall. It shows why the surface water in the county is generally quite hard.

8.6 (26.5) The reddish, sugar loaf-shaped hill on the skyline is part of the Jurassic Franciscan Formation, most likely a mass of altered iron and magnesium-rich igneous rock called a dunite. Dunites in the Franciscan are blocks of old oceanic crust that were plastered and subducted beneath the leading edge of the continent in Jurassic and Cretaceous time, about 60-100 million years ago.

8.9 (25.3)

to 9.8 (26.8) The road climbs out of Happy Canyon across the Paso Robles Formation by means of a series of switchbacks. The uppermost switchback at the top is on the South Branch of the Little Pine fault. Although the fault itself is not obvious, its presence is very evident because of the abrupt change in rock type from weakly consolidated Paso Robles Formation to the distinctly different, dark-colored shales of the marine Cretaceous Espada Formation. The great difference in age of these two rock units alerts geologists to the possibility of a fault, though neither the change in rock type nor the great disparity in age alone is adequate to prove the existence of a fault. It is also necessary to find actual evidence of displacement along the contact between these rocks. This often requires careful mapping and searching for an exposure of the fault plane in a creek bank or a hillside.

10.0 (25.1) The blue-green rocks in the hillside on the left (west) are serpentinites of the Franciscan Formation.

10.5 (24.6) Two prominent peaks can be seen on the right (east) in the distance, the peak on the left (north), at about 2 o'clock, is McKinley Mountain (6182 ft), and the other at 3 o'clock, is Santa Cruz Peak (5570 ft). Both are developed on marine late Cretaceous sandstone, the Cachuma Formation.

10.7 (24.4) You cross the north branch of the Little Pine fault here, leaving the Paso Robles Formation and crossing onto the serpentinites of the contrasting Franciscan Formation.

11.1 (24.0) The road crosses a landslide here. Several more landslides occur in the next 1.3 miles until reaching the Camuesa fault. The landslides are recognized by the typical hummocky surface and often result in a jumble of rock types. The slippery serpentinites of the Franciscan Formation are notably unstable and result in frequent landsliding wherever they occur.

11.7 (23.4) Just downhill to the right (east) is a large block of Franciscan serpentinite 20 to 30 feet in diameter surrounded

	by softer, less coherent rock. This sort of large isolated block is very common in Franciscan terrain and is called a "knocker".
12.4 (22.7)	Cross the Camuesa fault. Note the crushed, rubbly rock exposed in the road cut on the left (west). This is fault gouge or breccia.
12.5 (22.6)	Well-bedded Monterey cherty shale here.
12.9 (22.2)	Cachuma Forest Service Campground. Cachuma Creek and the road follow the trace of the Cachuma fault for a little less than half a mile before reaching an exposure of the marine Espada Formation of Cretaceous age. This rock is chiefly a dark-colored shale with some minor sandstone beds.
14.2 (20.9)	The steeply inclined rusty-colored rocks forming the ridge on the left (west) are examples of altered diabase of the Franciscan Formation. These rocks were lavas erupted in deep oceanic waters. Diabases have a composition very similar to basalt.
15.2 (19.9)	Cachuma Saddle and Forest Service Guard Station. The former Red Rock Quicksilver Mine is located about half a mile southeast of the saddle. Rocks here belong to the Cretaceous Espada Formation. Take Figueroa Mountain Road to the left (west).
15.8 (19.8)	Serpentinite is exposed in the road cut here.
16.2 (18.9)	Cross the Cachuma fault here and enter an exposure of Monterey Formation.
18.2 (16.9)	A good exposure of cherty Monterey shale in the road cut.
18.9 (16.2)	Ranger Peak (4652 ft) is uphill to the left (south). Its summit, like many other prominent peaks in the San Rafael Mountains, is formed from Monterey Formation.
19.4 (15.7)	Cross the Camuesa fault here, leaving the Monterey Formation and crossing onto the Franciscan serpentinite.
20.2 (14.9)	Good exposure of shiny serpentinite in the road cut on the right (north).
20.8 (14.3)	Good exposure of reddish Franciscan radiolarian chert in the road cut on the right (north). Cross Camuesa fault here. In the next half mile, the road crosses three strands of the Camuesa fault.
21.8 (13.3)	Figueroa Forest Service Campground.

22.2 (12.9)	Monterey Formation exposed on the right (north).
22.7 (12.4)	Junction with Catway Road. North of Figueroa Mountain, along Catway Road about 2.5 miles from this junction, is a small exposure of submarine volcanic rock showing pillow structure. Marine sedimentary rocks occur both above and below these volcanic rocks, which have been dated between 17 to 20 million years. They were erupted about the same time as the Tranquillon volcanic rocks in the southwestern part of the county, and the Obispo Tuff that occurs near Santa Maria.
23.2 (11.9)	Figueroa Mountain Forest Service Guard Station.
23.4 (11.7)	Green serpentinite of the Franciscan Formation is present on the right (north).
23.7 (11.4)	The pyramidal peak at 3 o'clock on the right (north) is Zaca Peak (4341 ft). It is on Monterey Formation and lies a short distance south of Zaca Lake.
26.7 (8.4)	Cross the poorly exposed Little Pine fault. However, the topography uphill from the fault is developed in the Franciscan Formation and is notably more irregular than the younger Paso Robles downhill from the fault. Irregular knockers composed of chert, greenstone and serpentinite mark the presence of Franciscan rock. Patches of trees and distinctive plants with intervening barren areas show evidence of plant toxicity resulting from the excess of magnesium in Franciscan soils. The Paso Robles Formation downhill is usually grass-covered.
28.2 (6.9)	After several tight switchbacks and steep grades, the road descends to the valley of Alamo Pintado Creek, which it follows to the town of Los Olivos and State Highway 154. The road hugs the northwest side of the valley to the junction with Sycamore Canyon, where it crosses to the southeast side. The uplands along the road are all underlain by Paso Robles Formation and the valley floors are nearly level, modern stream deposits (alluvium).

34.0 (1.1)   Good exposure of the Paso Robles Formation on the right (west).

35.1 (0.0)   Junction with State Highway 154 in the town of Los Olivos.

Santa Maria, Sisquoc, Foxen Canyon, Los Olivos, Ballard Canyon, Solvang, and Nojoqui Falls

This road provides something of a cross-section of the central part of the county. It extends from the agricultural lands of the Santa Maria Valley, across oil fields and the foothills of the San Rafael Mountains, to the foot of the north face of the Santa Ynez Range at Nojoqui Falls County Park.

0.0 (46.8)   Junction of Betteravia Road and U.S. Highway 101. The road here is on the older dune sand, the intermediate dune sand between the Orcutt sand and modern dunes. Turn right (east) at the top of the offramp.

0.7 (46.1)   Cross onto probable older alluvium. The contact is very indistinct here.

2.4 (44.4)   Leave the older alluvium and descend onto the Quaternary alluvial deposits of the Santa Maria River.

3.9 (42.9)   Begin wide right turn onto Foxen Canyon Road.

4.8 (42.0)   Begin wide left turn at the junction with Dominion Road. The higher surface to the right (south) is developed on the older alluvium.

6.5 (40.3)   Fugler Point. The Careaga sand is capped here by older alluvium. The prominent bluffs to the left (east), across the river channel are in San Luis Obispo County and are exposures of the middle Miocene Obispo tuff.

7.5 (39.1)   Village of Garey.

8.2 (38.6)   Junction with Orcutt Road. The hills to the right with oil wells are composed of older dune sand. The wells are part of the Santa Maria Valley Oil Field. Oil is derived, for the most part from Miocene rocks, mainly the Monterey Formation.

10.0 (36.8)   Town center of Sisquoc. Continue on Foxen Canyon Road.

10.9 (35.9)   Sisquoc sand and gravel plant on the left (north). These materials are derived from Sisquoc River gravel.

11.0 (35.8)   Follow Foxen Canyon Road to the right (south).

11.4 (35.4)	Follow Foxen Canyon Road to the left (east).
12.5 (34.3)	Junction with Tepusquet Road. Foxen Canyon Road lies here at the base of a bluff made up of Paso Robles Formation capped with older alluvium from the Sisquoc River.
13.5 (33.3)	Turn right at the old Sisquoc church. The road closely follows the base of the bluff on the right (west), composed of Paso Robles Formation, but the road itself is on the alluvium of Foxen Canyon Creek. Exposures of Paso Robles can be seen in the road cuts on the right (west) for the next mile or so.
15.3 (31.5)	Rancho Tinaquaic.
18.0 (28.8)	The higher hills on the right (southwest) are mostly composed of Sisquoc Formation; those on the left (east), mostly of the Monterey shale, dipping gently under the younger Sisquoc.
21.2 (25.6)	Junction with Alisos Canyon Road. Hills on the right (south) are mostly marine Careaga sand capped by Paso Robles Formation. Low hills to the left (east) are also Paso Robles.
22.9 (23.9)	Begin ascent onto the Paso Robles Formation from stream alluvium.
23.7 (23.1)	Junction with Zaca Lake Road to the left (east). The Foxen Canyon Road climbs onto the older alluvium that caps the Paso Robles Formation here.
27.2 (19.6)	Turn left (southeast) on Foxen Canyon Road. Zaca Station Road continues straight ahead to U.S. Highway 101.
28.4 (18.4)	The Paso Robles Formation is exposed in the road cut on the left (east). Good examples of this formation can be seen to the left (east). Well-developed river terraces along Zaca Creek can be seen to the left (southeast).
31.6 (15.2)	Cross State Highway 154 at Los Olivos, turn sharp right (west) immediately, on Ballard Canyon Road and climb by switchbacks onto the Paso Robles Formation.
33.1 (13.7)	Small but obvious slump on the hillside to the left (east), in the older alluvium.
34.3 (12.5)	Descend onto the alluvium of Ballard Creek. The little hill on the right (west) where the road bends to the left (east) is composed of Monterey Formation.

38.3 (8.5)	Junction with State Highway 246 in the center of the town of Solvang. Turn left (east). Turn right (south) at the traffic signal on Alisal Road. The town of Solvang is located, for the most part, on an old, high-level stream terrace deposited by the Santa Ynez River.
39.0 (7.8)	Descend onto a younger stream terrace on the north side of the Santa Ynez River channel.
39.3 (7.5)	Santa Ynez River bridge. Tranquillon volcanic rocks are present on either side of the road at the south abutment of the bridge, but these are not readily seen from the road.
40.2 (6.6)	Most of the rocks exposed in road cuts on the right (west) for the next mile belong to the early Miocene marine Vaqueros sandstone.
43.0 (3.8)	For the next half mile, the road crosses poorly exposed Sespe conglomerate of Oligocene age.
43.5 (3.3)	Descend onto the alluvium of Nojoqui Creek.
45.0 (1.8)	Junction of entrance road to Nojoqui Falls County Park. The falls are near the contact of the Eocene Anita shale and the Cretaceous Jalama sandstone. Where it falls over the bedrock, the water has deposited travertine (calcium carbonate or lime). Algae often contribute to the formation of travertine by extracting lime from the water. The falls are about 100 feet high.
45.8 (1.0)	Turn right (north) on Old Coast Highway (State Highway 1). If you are following this log from the north, begin log at this point, and subtract 1.0 mile from all readings.
46.3 (0.5)	Bridge across Nojoqui Creek. Sespe conglomerate on the right (east).
46.8 (0.0)	Junction with U.S. Highway 101.

Cat Canyon Road from Sisquoc via Palmer Road to U.S. Highway 101

This route crosses the main part of the Cat Canyon Oil Field and displays only the Miocene and younger rocks of Santa Barbara County.

0.0 (11.1)	Junction of Cat Canyon and Palmer roads in the village of Sisquoc. The road follows the valley floor on young stream

	alluvium of Cat Canyon. The hillsides are Paso Robles Formation with a capping here and there of older alluvium. As elsewhere, the Paso Robles Formation is weakly cemented and contains abundant chips from the Monterey Formation.
1.6 (9.5)	Pass through the Texaco portion of the Cat Canyon Oil Field.
2.0 (9.1)	Junction of Foxen Canyon and Cat Canyon roads. Turn left (south) on Cat Canyon Road.
3.5 (7.6)	The hills on either side of the road for about the next 2 miles are composed of the marine Careaga sand of Pleistocene age.
6.5 (4.6)	Shortly after the road turns right and leaves the alluviated floor of Cat Canyon, it climbs, by a series of switchbacks across an exposure of the marine Sisquoc Formation of Miocene age (visible in road cuts). The Sisquoc here is a thin-bedded diatomaceous shale and is the oldest rock seen on this trip. As the summit is approached, the road crosses onto the Pliocene Careaga sand and as the road descends beyond into the head of Howard Canyon, the Sisquoc again appears at the foot of the hill.
10.0 (1.1)	The rest of the way to the junction with U.S. Highway 101, the road is on the stream alluvium of Howard Canyon, with hills on either side composed of Paso Robles Formation.
11.1 (0.0)	Junction with U.S. Highway 101.

Drum Canyon Road from Los Alamos to State Highway 246

This route provides some good exposures of the Sisquoc Formation of late Miocene age as well as the marine Pliocene Careaga sand.

0.0 (9.4)	Junction of State Highway 135 and Centennial Avenue in the center of the town of Los Alamos. Take Centennial Road south through town.
0.5 (8.9)	The foothills here are composed of Paso Robles Formation.
0.9 (8.5)	Cemetery. A belt of Careaga sand crosses the road here and for the next 0.2 miles. It is not at all well exposed, but the valley followed by the road, Cañada de las Calaveras, widens appreciably as it crosses the soft and easily eroded Careaga.
1.0 (8.4)	The valley narrows as the road crosses onto the marine Sisquoc

Formation, which is present in the hills on either side of the road. The Sisquoc, though hardly a resistant rock unit, is still more resistant than the overlying, younger Careaga sand.

2.5 (6.9) For the next 3.5 miles, there are numerous exposures of the late Miocene Sisquoc Formation.

6.9 (2.5) The Pliocene marine Foxen claystone is exposed in the lower walls of the small canyon entering the valley of Santa Rosa Creek from the left (east).

7.1 (2.3) The hills on both sides of the valley of Santa Rosa Creek are composed of the Careaga sand. There is a good exposure of this rock across the creek to the left (east).

7.7 (1.7) A small sandpit on the right (west) exploits the Careaga sand. The hills on either side of the road beyond the sandpit are Paso Robles Formation with a cover of older alluvium.

9.4 (0.0) Junction with State Highway 246.

State Highway 135 (San Antonio Road) From Los Alamos to State Highway 1 via Vandenberg Air Force Base and Casmalia

This road passes through attractive agricultural crop land in the valley of San Antonio Creek and by one of the few inland marshy areas in the county-Barka Slough. The road crosses the Casmalia Oil Field, passes close to the former Airox Mine and ends on the Orcutt sand southwest of Santa Maria.

0.0 (24.3) Begin route at the corner of Centennial Avenue and State Highway 135 in the center of the town of Los Alamos. Follow State Highway 135 west (straight ahead). Los Alamos is situated on the alluvium of San Antonio Creek. For the first 9 miles the axis of the valley more or less follows the trace of the Los Alamos syncline.

5.0 (19.3) The flat, alluviated floor of the valley of San Antonio Creek is largely used for vegetable crops and some grapes. The low hills on either side of the valley are formed from the Paso Robles Formation.

6.5 (17.8) Bridge across San Antonio Creek.

8.9 (15.4) Turn left off of State Highway 135 at the junction with West San Antonio Road, continuing west. The hills west of this junction are capped by a sheet of Orcutt sand, which also is

	present along the road on the left (south) side after it crosses the valley.
9.9 (14.4)	Orcutt sand is exposed in the road cuts on the left (south) side of the highway. The poorly-drained valley floor here, on the right (north) is known as Barka Slough. It extends westward as far as the junction with State Highway 1, and is probably due to small-scale folding in the nearby Los Alamos syncline, resulting in a minor sag in this part of the valley of San Antonio Creek, thus ponding the stream slightly.
12.9 (15.4)	At the junction with State Highway 1 turn left (southwest). The base of the hills to the left (south) are made up of Foxen mudstone, a marine unit of middle to late Pliocene age, slightly older than the Careaga sand.
13.4 (10.9)	Turn right (west) on West San Antonio Road, leaving State Highway 1. From here to the end of West San Antonio Road at the Lompoc-Casmalia Road, the Sisquoc Formation is exposed on the steep slopes on either side of the valley, and capped by a sheet of Orcutt sand.
16.2 (8.1)	Junction with the Lompoc-Casmalia Road. Turn right (north).
18.7 (5.6)	Good exposures of the Sisquoc Formation can be seen in road cuts here.
18.9 (5.4)	Junction with Bishop Road to the right (east).
19.3 (5.0)	Whitish Sisquoc Formation exposed in road cuts and here capped by Orcutt sand. This relationship persists for several miles until the road descends onto the floor of Shuman Canyon at Casmalia.
20.8 (3.5)	Junction with Point Sal Road. This road leads it into the town center of Casmalia, but a short distance beyond it is blocked by a gate at the Vandenberg Air Force Base boundary. It was formerly the easiest way to reach Point Sal.
22.2 (2.1)	Cross Casmalia Oil Field. Rocks on either side of the road belong to the Sisquoc Formation.
22.8 (1.5)	Road cuts in this vicinity clearly display Sisquoc Formation.
23.0 (1.3)	Airox Mine facility on the right (east). This operation once produced a light-weight, bubbly material by roasting an oil-soaked unit of the Sisquoc Formation. The product was used

chiefly as a lightweight aggregate for specialty concretes and for absorbent materials such as kitty litter. It was closed a few years ago because of air pollution problems associated with the release of sulfur dioxide gas.

23.3 (1.0) Good exposures of the Sisquoc Formation in road cuts.

24.3 (0.0) Junction with State Highway 1 at Black Road.

Clark Avenue and Dominion and Palmer Road Loop

This short trip crosses parts of both the Santa Maria Valley and West Cat Canyon oil fields. All the rocks seen on this trip are Pliocene or younger and most are land-laid deposits.

0.0 (10.4) Start the trip at the Junction of U.S. Highway 101 and Clark Avenue. Take Clark Avenue to the right (east). The road begins on the gently rolling surface of the Pleistocene Orcutt sand, the older of the three sand dunes in this part of the county.

2.8 (7.6) Turn right (south) on Dominion Road. Scattered oil wells are pumping from the Santa Maria Valley Oil Field. Road remains on the Orcutt sand of Pleistocene age.

4.1 (6.3) Good exposures of Orcutt sand in road cuts.

4.4 (6.0) Numerous oil wells of the West Cat Canyon Oil Field.

4.9 (5.5) Good examples of badland erosion in the road cuts where the soft Orcutt sand has been gullied by rain.

5.8 (4.6) Road leaves the Orcutt sand here and passes onto the older Paso Robles Formation, that underlies the Orcutt.

6.3 (4.1) Junction with Palmer Road. Turn right (south).

7.4 (3.0) Cat Canyon Oil Field pumping facility.

8.8 (1.6) As the road descends off the rolling surface of the Paso Robles Formation onto the flat floor of the stream valley, it crosses a narrow band, about 0.1 mile wide, of Careaga sand, a Pliocene marine deposit beneath the Paso Robles Formation.

10.4 (0.0) Junction with U.S. Highway 101. Road is on a young stream alluvium here.

Harris Grade Route: Clark Avenue and U.S. Highway 101 via Orcutt to Lompoc

This road is one of the two or three best routes in the county to see the details of the diatomaceous Sisquoc Formation of late Miocene age. Traffic is usually light over most of the route and one can stop safely at many road cuts.

0.0 (18.0)	Junction of Clark Avenue and U.S. Highway 101 south of Santa Maria. The road crosses the older dune sands that have gently rolling topography. Where exposed, the sand has a pale brown color because of iron-staining on the sand grains.
2.3 (15.7)	Turn left (south) on to State Highway 135 in the town of Orcutt. This junction is on stream alluvium deposited by Orcutt Creek. The slightly higher ground on either side of the road is the older dune sand.
3.3 (14.7)	Enter Graciosa Canyon and merge with State Highway 1. The hills to the right (west) are capped with marine Pliocene Careaga sand. Downslope, but younger, and covering the Careaga is the Paso Robles Formation and near the base of the hills and still younger is the Orcutt sand.
5.7 (12.3)	The divide between the north-draining Graciosa Canyon and the Harris Canyon drainage to the south is marked by exposures of Careaga sand on both sides of the road. The hill tops are capped with Paso Robles Formation.
6.7 (11.3)	Junction with Graciosa Road.
7.2 (10.8)	State Highway 1 veers to the right (west). Keep straight ahead.
10.1 (7.9)	Junction with State Highway 135 (San Antonio Road) to the town of Los Alamos. Keep straight ahead here. The Los Alamos Valley to the left (east) is drained by San Antonio Creek that reaches the sea on the Vandenberg Air Force Base 4 miles north of Purisima Point. The road junction is on alluvial deposits of San Antonio Creek.
10.9 (7.1)	The road climbs onto the Paso Robles Formation, exposures of which can be seen in road cuts on the left (east).
11.7 (6.3)	The Careaga Formation is displayed in the road cuts on the left (east).

12.3 (5.7) The Sisquoc Formation is exposed in road cuts on the right (west), and for the next 2 miles there are a number of excellent exposures of the white, highly diatomaceous Sisquoc Formation. This is certainly one of the best places in the county to see good exposures of this rock unit. Sweeney Road east of the town of Lompoc is another. Both the nearly pure diatomites and the more clay-rich shaly phases of the formation are well-displayed on the Harris Grade as it crosses the Purisima Hills.

12.9 (5.1) Good exposure of the Sisquoc Formation on the right (west).

13.4 (4.6) Oil wells in the Lompoc Oil Field. This field was discovered in 1903 by drilling wells near some tar seeps in Purisima Canyon about 2 miles east of the Harris Grade. The oil comes from the Monterey Formation at a depth of about 2000 to 3000 feet.

14.0 (4.0) Excellent exposures of the Sisquoc Formation diatomites in road cuts in this vicinity.

15.4 (2.6) Junction with Rucker Road. The road here is on Orcutt sand.

17.0 (1.0) Junction with Burton Mesa Road.

18.0 (0.0) Junction with State Highway 1 from the right (west).

Across the Sierra Madre via Colson and La Brea Canyons and Miranda Pine Mountain

This route provides a good cross-section of the rocks in the northwestern part of the Sierra Madre. Near Miranda Pine Mountain, a graded road to the southeast follows the crest of the Sierra Madre for about 20 miles to McPherson Peak (5749 ft). Although this crest road affords numerous views of the upper Sisquoc River Valley and the Cuyama Valley, the rocks exposed along the road are a fairly monotonous series of late Cretaceous and Paleocene marine sandstones.

0.0 (34.7) Junction of Tepusquet and Colson Canyon roads. Rocks on both sides of the road are typical cherty Monterey shale. The road crosses Colson Creek a number of times en route to the divide between Colson and La Brea canyons.

0.7 (34.0) Exposure of the fine-grained Coast Range ophiolite on the right (south). This rock is an altered sea floor eruptive rock called a diabase.

1.7 (33.0)	The high, prominent ridge ahead is a massive dolomite bed in the Monterey Formation. Dolomites are carbonate rocks much like limestones, but have higher content of magnesium.
2.5 (32.2)	A good example of the Monterey Formation on the right (south).
4.1 (30.6)	Colson Forest Service Campground.
4.7 (30.0)	Gate just beyond the cattle guard. The East Huasna fault crosses the road near the summit, but is not well exposed because here the Monterey Formation is on both sides of the fault. The road descends into La Brea Canyon via several switchbacks across Monterey shale.
7.8 (26.9)	Just before reaching the valley floor, the road crosses lower Monterey Formation rocks and onto a band of sandstone called the Hurricane Deck unit, probably equivalent to the Vaqueros Formation. Just before reaching the floor of the valley the road crosses the Cretaceous sandstone of the Cachuma Formation.
7.9 (26.8)	At the junction with La Brea Canyon Road turn left and head north. The hillsides on both sides of the road are made up of the late Cretaceous marine Cachuma Formation, mainly sandstones, but with numerous wide bands of shale. As a rule, the road surface is much rougher across the sandstones, but smoother and slicker after rains, across the shales.
12.8 (21.9)	Wagon Flat Forest Service Campground. The trail to the east to Lazy campground continues to the old White Elephant Barite Mine. All the rocks here are part of the Cachuma Formation.
13.7 (21.0)	Gate.
14.7 (20.0)	A prominent sandstone bed forms a cliff at the ridge top to the right (east). This is part of the Cachuma Formation.
15.0 (19.7)	An exposure of Cachuma Formation shale.
15.5 (19.2)	Junction with the Buckhorn offroad vehicle trail on the left (west).
16.5 (18.2)	Junction with Pine Canyon Road on the left (west).

18.1 (16.6)	Small patch of possible red, nonmarine Oligocene Simmler Formation on the left (west). This rock may also represent a nonmarine phase of the Cachuma Formation, however.
22.5 (12.2)	Pine Flat area on the left (north). This is the approximate position of the major Rinconada fault, a unit of the Sur-Nacimiento fault zone that forms the western boundary of the Salinian block with its granitic basement. The Rinconada fault crosses State Highway 166 near Clear Creek.
24.0 (10.7)	Excellent exposures of pebble conglomerate on the left (west). These conglomerates are a phase of a late Cretaceous and Paleocene marine sequence. A formation name is not yet formalized.
25.8 (8.9)	A four-way junction. Sierra Madre Road continues both to the southeast and northwest. The road to Miranda Pine Forest Service Campground goes to the right (north) up the hill to Miranda Pine Mountain (4061 ft). Follow Sierra Madre Road to the northwest.
29.6 (5.1)	Junction with Old Sierra Madre Road to the left (northwest). Locked gate 2.5 miles beyond this junction on Old Sierra Madre Road. Keep right (north).
30.7 (4.0)	A fine view of the upper Cuyama River Valley. Note the meander beds in the river course.
31.2 (3.5)	Good exposure of the probably Paleocene conglomerate.
32.3 (2.4)	Red rocks on the left (west) are nonmarine, unfossiliferous and apparently equivalent in age to the Cretaceous Cachuma Formation.
32.8 (1.9)	Good exposure of the Paleocene and earliest Eocene marine conglomerate on the left (west).
32.9 (1.8)	Good view of the broad upper Cuyama River Valley.
33.7 (1.0)	Paleocene and earliest Eocene thin-bedded sandstone on the left (west).
34.7 (0.0)	Junction with State Highway 166.

Santa Rosa Road from U.S. Highway 101 at Buellton to State Highway 1 near Lompoc

This road follows the south bank of the Santa Ynez River from the Santa Ynez Valley at Buellton, through the narrows between the Santa Rita and Santa Rosa hills to the eastern end of the Lompoc Valley. It affords many views of the river as well as a considerable variety of rock units in the Santa Rosa Hills on the left (south).

0.0 (16.7) Offramp on U.S. Highway 101 to the town of Buellton and Santa Rosa Road. Turn left at head of ramp. The road here is an old, elevated Santa Ynez River terrace.

0.2 (16.5) Turn left on Santa Rosa Road (south).

0.7 (16.0) The oak-covered hillside on the left (south) is made up of Sespe Formation conglomerate, an old river gravel of Oligocene age.

0.9 (15.8) The light-colored sandstone in the hillside to the left (south) is the Vaqueros Formation of latest Oligocene to early Miocene age, deposited in a shallow water marine environment, on top of the nonmarine Sespe.

1.5 (15.2) The high cliffs on the left (south) are exposures of the Vaqueros Formation.

1.8 (14.9) The extensive grass-covered hillside on the left (south) is underlain by the Rincon mudstone. This rock rests on the older Vaqueros and weathers to a clay-rich, adobe-like soil that generally supports grass rather than trees or chaparral.

2.2 (14.5) The oak-covered hills straight head are underlain by volcanic rocks chiefly basaltic flows of the Tranquillon volcanic Formation.

2.7 (14.0) An exposure of this weathered volcanic rock is next to the road on the left (south). The main, sandy channel of the Santa Ynez River is close to the road on the right (north).

3.9 (12.8) Oak woodland on the Vaqueros Formation on the left (east).

4.3 (2.4) The clearly layered rocks in the road cut on the left (south) belong to the Gaviota-Sacate Formation. These two units are here not readily distinguished from one another. They are of Eocene to Oligocene age.

5.2 (11.5)	The rolling grass-covered hills with scattered oaks are underlain by the Gaviota-Sacate Formation, a marine rock. As noted above, the two formations grade into one another in this area, though they are clearly separable at most localities on the south face of the Santa Ynez Mountains.
6.5 (10.5) to 7.9 (9.2)	Pebbly older stream gravels deposited by the Santa Ynez River can be seen in the road cut on the left (south). The higher, grass-covered hills beyond are developed on the Rincon mudstone.
6.9 (9.8)	Note the flat surfaces of the old river terraces on the far side of the river channel. These are abandoned river-deposited benches left by the river as it cut its channel downward. Sometimes several such benches can be seen at a given locality.
7.3 (9.4)	The steep hill on the left (south) is underlain by the Monterey Formation that is exposed in the road cut.
7.5 (8.8)	Most of the hummocky surface of the hillside to the left (south) is the result of an old landslide composed of deformed, crumpled and broken Monterey cherty shale.
9.8 (6.9) to 14.9 (2.0)	Entrance to Rinconada Winery. Distant oak-covered hills on the left (south) are developed on the Monterey Formation.
10.3 (6.4)	There is a good exposure of the Monterey Formation in the road cut on the left (south).
13.8 (2.9)	More good exposures of the Monterey shale in the road cut on the left (south).
14.2 (2.5)	Well-bedded diatomaceous Sisquoc Formation is exposed in the road cut on the left (south).
14.7 (1.8)	Good exposures of the pale colored, cherty Monterey Formation in road cuts on the left (south).
16.2 (0.5)	Salsipuedes Creek bridge.
16.7 (0.0)	Junction of Santa Rosa Road and State Highway 1.

Sweeney Road from State Highway 246 to End of the Road

This road and the Harris Grade offer the best views of the highly diatomaceous parts of the Sisquoc Formation that are readily accessible, though purer diatomites occur at the Celite and Grefco quarries to the south.

0.0 (5.7)	Junction of Sweeney Road and State Highway 246. A county road maintenance yard is located on the right (west) of this junction. Sweeney Road follows the north side of the strongly meandering path of the Santa Ynez River across an anticlinal fold in the rocks. This folding and perhaps additional uplift have caused the river to downcut its channel, preserving the meandering pattern which was inherited from a time when the river occupied a broader, more open valley.
0.5 (5.2)	Pebbles exposed in the road cut on the left (north) were deposited by the river, but subsequently raised well above the present stream.
1.0 (4.7)	Badland gullying is developed in an exposure of Orcutt sand, a Pleistocene dune deposit.
1.1 (4.6)	This road cut shows Orcutt sand resting on whitish Sisquoc diatomaceous shale.
1.5 (4.2)	Excellent exposure of the Monterey Formation in the hillside to the left (north).
1.8 (3.9) to 5.2 (0.8)	Cross the axis of a small synclinal fold. The folding can be seen in the road cut. Exposed rocks belong to the Monterey Formation.
2.0 (3.7)	Excellent exposure of the Monterey Formation in the road cut.
2.4 (3.3)	A nicely exposed anticlinal fold is displayed in the river-cut cliff on the left (east).
4.9 (0.5)	Excellent exposures of Sisquoc diatomaceous shales in road cuts on the left (north).
5.7 (0.0)	End of road. Site is on a young river terrace.

Tepusquet Canyon from the Cuyama River Road to the Sisquoc River

This route cuts across the northwestern end of the San Rafael Mountains. On the north, rocks are sedimentary marine deposits of Cretaceous age; rocks on the south are mainly the familiar Miocene Monterey Formation. About half way, there is an example of what is called a "diapir", in which serpentinite has been squeezed upward through the Monterey Formation.

0.0 (19.1)   Junction of Tepusquet Road and State Highway 166 (Cuyama

River Road). Rocks exposed on the first part of this road are marine deposits of late Cretaceous age, assigned by some geologists to the Cachuma Formation; by others to the Morris and Buckhorn formations.

3.0 (16.1) A good exposure of Cretaceous sandstone on the right (west).

3.6 (15.5) Reddish rocks exposed for the next 0.25 mile have been called Sespe Formation by some geologists, the Simmler Formation by others. All agree that this deposit is most likely of Oligocene age and that it represents land-laid stream and flood plain deposits.

4.1 (15.0) The rocks on the left (east) are called Vaqueros sandstones by some geologists, or Hurricane Deck sandstones by others. The age is early Miocene; they formed as the sea spread over the floor plains of the preceding "Sespe" Formation.

5.1 (13.9) Summit.

8.7 (10.4) The Monterey Formation is well exposed on the right (west).

9.1 (10.0) Just beyond the mouth of Suey Canyon entering from the right (northwest), the road crosses a patch of altered volcanic rock, originally a diabase, but now serpentinite. There are numerous landslides on the right and exposures are not very good, but this is an interesting example of what is called a cold intrusion, or diapir, in which a mass of Jurassic serpentinite was squeezed upward through the Monterey Formation. Similar, but much larger examples occur elsewhere in the Coast Ranges, notably at Mount Diablo in Contra Costa County and at New Idria in San Benito County.

10.2 (8.9) Junction with Colson Canyon Road to the left (east). The Colson Canyon Road climbs over a ridge composed of Monterey Formation and descends into La Brea Canyon, whose bedrock is the Cretaceous Cachuma Formation (see Colson Canyon log).

12.6 (6.5) Good exposures of Monterey shale on the left (east).

14.7 (4.6) Junction with Santa Maria Mesa Road. Turn right here. Tepusquet Road does continue straight ahead, but in recent times has been closed at the Sisquoc River crossing. The Santa Maria Mesa Road crosses young river deposits of the Sisquoc

River all the way to the junction with Foxen Canyon Road just north of Garey.

18.5 (0.6) Sisquoc River bridge.

19.1 (0.0) Junction with Foxen Canyon Road.

Corralitos Canyon (Brown Road) State Highway 1 to Point Sal Beach

This road may be closed at the Corralitos Ranch because of storm damage and washouts on the steep, seaward side of Point Sal Ridge. Part of the road crosses Vandenberg Air Force Base property and apparently neither the County of Santa Barbara nor the Air Force is interested in maintaining it. A great pity because the geology of the Point Sal Ridge is of unusual interest.

0.0 (8.4) Junction of Brown Road and State Highway 1.

1.4 (7.0) The whitish rock straight ahead is nearly vertical Sisquoc diatomaceous shale.

4.0 (4.4) Leave paved road at Corralitos Ranch entrance. Turn right (west).

4.2 (4.2) Start up the grade and cross onto the red, nonmarine Lospe Formation. This rock resembles the Sespe Formation found at many other places in the county, but it is younger than most of the Sespe.

5.0 (3.4) For about the next 0.75 of a mile the road crosses weathered and poorly-exposed submarine basalt flows. Exposure of this same rock near Point Sal is much fresher and the pillow structure can easily be seen (Figure 37). These lavas form the bedrock along the road on both sides of the ridge crest. The road crosses onto the soft Point Sal shale as it descends the seaward slope of Point Sal Ridge. The road ends at a small parking area on a small landslide. The Point Sal Formation is a marine deposit of the middle Miocene age.

8.4 (0.0) Parking area. If you succeed in reaching the beach, the rocky headland on the left (southeast) is an exposure of Tertiary diabases, probably equivalent to the Tranquillon volcanic sequence near Point Arguello. At the western end of the beach is an ophiolite sequence that extends westward to Point Sal. These rocks represent an unusually complete ophiolite sequence formed on the deep sea floor during Jurassic time, and plastered against the continent during a period of subduction. The nature of an ophiolite sequence was discussed in the description of the Point Sal headland in Chapter 2.

Cuyama River Valley Route from Santa Maria to the Junction of State Highways 166 and 33

This is the only highway in the county that crosses most of the Southern Coast Ranges. Although State Highway 166 begins near Guadalupe west of Santa Maria, all that portion of the road lies on Quaternary alluvium deposited by the Sisquoc and Santa Maria rivers and there are no outcrops of bedrock to be seen. The most interesting features are the sand dunes that lie between Guadalupe and the coast and these are discussed elsewhere in this book.

State Highway 166 follows the course of the Cuyama River east of Santa Maria. Because the river marks the boundary between Santa Barbara and San Luis Obispo counties, the road crosses back and forth from one county to the other, and the comments in this log mention features on either side of the county line if they can be seen from the highway.

The road log begins at the Junction of State Highway 166 and U.S. Highway 101 in San Luis Obispo County just north of Santa Maria. The bluffs that can be seen from the Santa Maria bridge to the right (east) are composed of Orcutt sand capped by patches of old stream alluvium.

0.0 (73.5) At the junction, the road is on Orcutt sand.

2.5 (71.0) The craggy outcrops on both sides of the highway are exposures of the Miocene Obispo tuff, a volcaniclastic deposit possibly erupted from vents near San Luis Obispo and Morro Bay. There are a series of volcanic plugs in that vicinity, the most prominent of which is Morro Rock.

4.6 (68.9) Junction with Suey Road on the left (north).

6.3 (67.2) Turnout where one can view the Twitchell Reservoir. This reservoir was built for flood control and ground water replenishment. It controls the flow of the Cuyama River, but not the Sisquoc River, which joins the Cuyama drainage a short distance downstream from the dam.

7.2 (66.3) Good exposures of the siliceous shales of the Monterey Formation. Other good exposures for the next 5 miles.

Figure 73. Cuyama River near its junction with Cottonwood Creek; view east. The Caliente Range in the distance is in San Luis Obispo County. This range is composed mostly of Miocene sedimentary rocks.

8.9 (64.6) Alamo Creek bridge.

10.7 (62.8) Cross the axis of a prominent anticlinal fold exposed in the Monterey Formation on the left (north).

11.1 (62.4) Large road cut with an excellent exposure of Monterey Formation.

12.5 (61.0) Massive sandstones in the bluff to the left (north) are Miocene marine Vaqueros rocks resting on Cretaceous sandstones.

13.4 (60.1) Approximate location of the East Huasna fault. The road begins to follow the course of the Cuyama River at this point.

15.2 (58.3) Cross into Santa Barbara County at bridge. The junction with Tepusquet Road is just beyond. The highway cuts across exposures of the various units of the Franciscan Formation for about the next 5 miles. This patch of Franciscan rock is known as the Stanley Mountain window because younger rocks that formerly covered the Franciscan have been eroded

	away, exposing these older rocks below. Stanley Mountain (2475 ft) lies about 3.5 miles north of the highway in San Luis Obispo County.
16.4 (57.1)	A good exposure of Franciscan graywacke sandstone in the road cut on the right (south).
17.4 (56.1)	A nice exposure of bluish-green Franciscan serpentinite across the river on the left (north). Recall that serpentinites usually result from the metamorphism of volcanic rocks like basalt and diabase.
17.9 (55.6)	Exposure of Franciscan melange. This exposure has a large "knocker" of more resistant graywacke surrounded by sheared and crushed, softer, more yielding shale. Recall that a melange is a potpourri of various rock types mixed together, usually during the subduction process at the edge of a continent.
19.2 (54.3)	The high ridge straight ahead is an exposure of Jurassic ophiolite in the Franciscan, and derived from the alteration of dark-colored volcanic rocks such as basalt and diabase. This rock is exposed for about the next 0.5 mile in both road cuts and across the river in stream bluffs.
21.0 (52.5)	Upper Cretaceous sandstone and shale can be seen in road cuts on the right (south). These rocks rest on the older Franciscan Formation.
21.6 (51.9)	Bridge across the Cuyama River. Enter San Luis Obispo County.
22.5 (51.0)	The hills across the river to the left (north) are early Miocene conglomerates of the Vaqueros Formation.
24.5 (49.0)	Cross the Rinconada or Sur-Nacimiento fault zone. This is a major structural feature of the southern Coast Ranges and forms the southwestern boundary of the Salinian Block with its granitic basement. Near the northeastern end of the road cut, nonmarine Simmler conglomerate is separated from upper Cretaceous sandstone by a wide zone of crushed rock (gouge) marking the fault.
24.7 (48.8)	An exposure of the red Simmler Formation, here a conglomerate, can be seen on the right (south). The Simmler and other rocks northeast of the fault zone lie on Salinian granitic base-

ment, though none of the granitic basement is exposed in Santa Barbara County. It is, however, exposed not far north in San Luis Obispo County.

25.8 (47.7) Sierra Madre Road on the right (south). The South Cuyama fault crosses the highway here, but is not well exposed. This is a thrust fault along which the Sierra Madre has been raised and shoved northeast with respect to the Cuyama River Valley. The older rocks of the Sierra Madre, mainly Eocene and Paleocene sandstones, have been uplifted more than 6000 feet with respect to the Oligocene and Miocene Simmler Formation on the valley side of the fault.

27.6 (45.9) Leave the Cuyama River gorge and enter the broad, upper Cuyama Valley. Note the broad river terraces here and the meandering course of the river.

30.1 (43.4) Note the good examples of meander bends in the Cuyama River on the right (south).

32.2 (41.3) The highway here crosses the surface of an old river terrace. Some of the river terrace gravels can be seen in low road cuts.

34.0 (39.5) The prominent mountain range straight ahead on the northeast side of the valley is the Caliente Range, composed mostly of Miocene marine sedimentary rocks. The Caliente Range is bounded on the Cuyama Valley side by the Morales thrust fault, so the Cuyama Valley is caught between two mountain fronts that are being shoved toward one another. This has caused the intervening valley block to be folded downward into a syncline (Figure 73).

The Cuyama Valley is the location of two major oil fields as well as several minor ones. Most of these are now largely depleted. They were discovered in 1948-49 by Thomas Dibblee, Jr., when he was employed by the Richfield Oil Company.

36.4 (37.1) Horizontal sedimentary rocks on the left (north) are nonmarine Pliocene deposits.

38.4 (35.1) Miocene lake bed deposits can be seen across the river on the right (south). The low hills nearby show numerous small landslides or slumps.

38.9 (34.6) Cross Cuyama River and enter Santa Barbara County.

40.4 (33.1)	Note the well-developed river terrace on the right (south).
41.7 (31.8)	Morales Canyon can be seen in the Caliente Range on the left (north). Just east of the canyon mouth is Whiterock Bluff, formed from the Monterey Formation, capped by the white sandstones of the Santa Margarita Formation.
44.3 (29.2)	The Morales fault may be seen along the base of the Caliente Range, folded Monterey Formation is above the fault, and ancient lake beds of the Morales Formation are below.
44.4 (29.1)	Russell Ranch Oil Field. This was one of the two major oil fields in the Cuyama Valley.
52.1 (21.4)	Town of New Cuyama.
56.3 (17.2)	The town of "old" Cuyama. The broad valley here is floored by alluvial deposits of the Cuyama River.
57.8 (15.7)	Enter San Luis Obispo County.
59.9 (13.6)	The high peak to the right (southeast), at about 1 o'clock in the far distance, is Cerro Noroeste (Mount Abel), whose summit has an elevation of 8286 feet. It lies in northwestern Ventura County and is an exposure of Precambrian granitic rock.
62.5 (11.0)	Junction with State Highway 33. Turn right (south) toward the town of Ojai.
64.7 (8.8)	Cuyama badlands on the left (east). These are developed on Pliocene nonmarine rocks, and like other badland areas is characterized by soft, fairly impermeable rocks in an arid and semi-arid environment. Sparse vegetation does not protect the easily-eroded rock from rapid gullying during occasional heavy rains. This results in a network of closely-spaced, steep-sided gulleys which erode so rapidly that protective vegetation has difficulty in getting established.
65.3 (8.2)	Santa Barbara County line. Leave San Luis Obispo County.
65.8 (7.7)	Entrance to Santa Barbara Canyon at 2 o'clock on the right (southwest). The paved road up this canyon ends approximately due east of Fox Mountain (5165 ft) and just downstream from the trace of the Ozena thrust fault, which together with the South Cuyama fault to the northwest, makes the base of the uplifted Sierra Madre.

66.1 (7.4)	Junction with Ballinger Canyon Road to the left (east). This road leads into the Cuyama badlands.
69.9 (3.6)	The town of Ventucopa. Note the well-developed but small alluvial fans at the base of the Cuyama badland hills on the left (east). The red rocks in the middle distance on the right (west) belong to the Oligocene, nonmarine Simmler Formation.
71.3 (2.2)	Junction with Quatal Canyon Road to the left (east). This road also leads into the Cuyama badlands.
73.5 (0.0)	Ventura County line. Nonmarine tilted Tertiary deposits are well exposed in the river bluffs to the right (west). The high ridge in the distance to the right (southwest), is Cuyama Peak (5875 ft) where Eocene and Paleocene marine rocks are exposed.

GLOSSARY

agglomerate – A rock composed of angular blocks of rock, usually of volcanic origin. Typically cemented by finer materials. See also breccia.

andesite – A dark-colored volcanic rock characteristic of major mountain belts. Somewhat more siliceous than basalt.

antecedent stream – A stream whose course was established before a barrier was raised in its path, but which managed to maintain its course across the barrier.

anticline – A folded structure, circular or elongate, in which beds or layers dip away from its axis. Like an inverted canoe. Compare with **syncline**.

asphaltum – Viscous to solid tarry material, typically associated with petroleum occurrences and tar seeps. A black hydrocarbon. See also **pitch**.

badland – An area of intricately eroded steep-sided gullies, usually in arid or semi-arid regions with soft, somewhat impermeable rocks.

barite – A mineral composed of barium sulfate ($BaSO_4$)

basalt – A dark-colored, silica-poor volcanic rock characteristic of oceanic sea floors. Very fluid when molten.

basement or **basement rock** – Crystalline rocks that underlie the oldest sedimentary rocks in a region. Commonly the core of major mountain ranges.

beach cell – A stretch of coastline that includes all sources of beach sand as well as the means for disposing of excess sand, usually a submarine canyon.

beach drift – The slow downcoast movement of sand on the face of the beach, but not including sand moved by the longshore current.

bedding plane – The surface between sedimentary layers, usually the result of an abrupt change in the environment of deposition.

Big Bend – Refers to the change in direction from northwest-southeast to more nearly east-west where the San Andreas fault crosses the Transverse Range Province.

black sand – A sand rich in heavy minerals usually mostly magnetite and other iron-bearing minerals, but often including other high-density minerals such as garnet, gold and platinum. A natural placer deposit.

black smoker – A submarine vent on a sea floor spreading center, from which hot, highly mineralized waters are discharged. Often with a chimney-like structure composed of sulfides of various metals.

blue schist – A metamorphic rock containing the blue mineral glaucophane and indicative of the subduction process.

boulder pile mountains – Mountains dominated by stacks of massive rocks such as sandstone or granite.

breccia – A rock composed of cemented angular blocks (clasts) of rock. Indicates minimal transport of the blocks, in contrast with a conglomerate which is characterized by rounded pebbles demonstrating more transport before deposition. See also **agglomerate**.

chaparral – Woody scrub vegetation characteristic of Mediterranean climatic regions such as coastal California.

chert – A hard, dense siliceous sedimentary rock composed of very fine-grained chalcedony or opal.

chlorite – A greenish, micaceous mineral, usually containing some iron. Common in metamorphic rocks.

chromite – The principal ore of chromium, an iron-chromium oxide.

clastic – A rock composed of broken fragments of other rocks, sandstones and conglomerates.

coal oil – An obsolete name for kerosene, a liquid hydrocarbon.

cold intrusion – See diapir.

cold seep – A submarine spring yielding cold, mineralized water.

columnar section – An illustration of the sequence of rocks in a given area, arranged by age and showing relative thicknesses and composition.

conglomerate – A sedimentary rock consisting of cemented, rounded pebbles or boulders. Compare with breccia.

continental borderland – The offshore area of southern California between Point Conception and the northern part of Mexico, dominated by deep basins and elevated intervening blocks and islands.

continental drift – The slow movement of continental crustal blocks with respect to one another, presumably driven by convection currents in the mantle below the crust.

continental shelf – A submarine shelf surrounding continents, of varying width, but seldom more than 600 feet below sea level.

cordillera – A major chain of mountains.

cross-section – A diagram showing the rocks present below the surface along a given line. A side view of a slice of crust.

cuspate delta – A stream deposit at its mouth, triangular in shape, in which the stream channel terminates at the apex of the triangle. Usually characteristic of minor streams rather than large rivers.

dacite – A fine-grained volcanic rock, similar to andesite, but slightly poorer in calcium.

deformation – The result of stress applied to a rock. Usually results in faulting, folding or uplift.

diabase – An igneous rock similar in composition to basalt, but with crudely aligned, lath-shaped feldspar crystals.

diapir – A structural feature, usually in sedimentary rock, in which plastic rock material from depth has been squeezed, without melting, into or through the overlying rocks. Also called a **cold intrusion**.

diatom – A microscopic, single-celled green plant (alga) growing in salt or fresh waters. Diatoms secrete siliceous shell-like structures called **frustules**. May be very abundant in some environments.

diatomite – A sedimentary rock composed largely or entirely of diatom frustules. Also called diatomaceous earth or **Kieselguhr**.

dike – A tabular, volcanic intrusion that cuts across the enclosing rocks. Compare with **sill**.

dimension stone – Building stone quarried in blocks for various uses.

diorite – A plutonic igneous rock, similar to a granite, but darker-colored, less siliceous and richer in iron.

dip – The inclination of a rock layer, a fault or a joint surface, measured from horizontal.

dip slip – The displacement along a fault in the direction of the dip of the fault.

dip slope – An exposure of sedimentary rock in which the ground surface slopes in the same direction as the dip of the rock.

dunite – A plutonic igneous rock composed entirely or chiefly of the iron-magnesium mineral olivine. Characteristic of subduction zones. See also **peridotite**.

elasticity – The ability to recover totally from stress or strain.

embayment – Any indentation along a shoreline.

estuary – An inlet along the coastline, usually at a river mouth. Typically shallow and with a muddy bottom.

fanglomerate – A sedimentary rock consisting of poorly sorted clasts, some often quite large, characteristic of alluvial fans in arid and semi-arid regions.

fault – Any fracture in rock along which there has been obvious displacement.

Flandrian Transgression – The flooding of coastal low lands that accompanied melting of the last continental glaciers during the closing days of the Pleistocene Epoch.

flow – A sheet of volcanic rock that while molten, flowed across a land surface or the sea bottom.

folding – Bending, without fracturing, of any layered rock.

foraminiferans – A group of single-celled protozoans that usually secrete shells (tests) of calcium carbonate (lime). Some form tests from sand grains. Often very abundant and very useful in dating sedimentary rocks.

Formation – A mappable rock unit, often sedimentary, and composed of a distinctive rock type or a distinctive association of several rock types. Named, usually, for the locality where it was first described.

fracture – Any break in a rock. In minerals, any break that does not follow cleavage lines.

fringing reef – A coral reef close to a shoreline with little or no open water between the beach and the reef.

frustule – A shell-like structure secreted by diatoms. Often shaped like a pillbox, very porous and highly variable with intricate designs.

fumarole – See solfatera.

gabbro – A dark-colored, coarse-grained plutonic igneous rock. Low in silica and high in iron. The deep-seated equivalent of the volcanic rock basalt.

gas hydrate – A somewhat unstable crystalline mixture of water and methane gas. Also called **methane clathrate**.

glaucophane – A blue mineral often produced in subduction zones and indicative of metamorphism under low temperature, but very high pressure. Common in the Franciscan Formation. A sodium-rich amphibole mineral.

gouge – Crushed or milled rock formed by movement along a fault. May be very fine-grained or brecciated. Also called **pug** by the British.

graywacke – A dark-colored, fine-grained sandstone often rich in volcanic materials. Typically formed by turbidity currents in deep water. Very common in the Franciscan Formation.

greenstone – Any hard, greenish, fine-grained metamorphic rock derived from igneous rocks such as basalt.

gusher – An uncontrolled fountain of petroleum, natural gas and water fed by a high-pressure oil reservoir. Rare today, but quite common in the early days of petroleum production.

hard water – Any natural fresh water containing an appreciable content of calcium and magnesium carbonates. Most surface and well waters in Santa Barbara County are hard.

hematite – A very common, red or black iron oxide mineral responsible for the color of red sedimentary rocks. One of the commonest of mineral pigments.

homocline – A sequence of sedimentary rocks that all dip in the same or similar direction.

hopper dredge – A scow designed to discharge dredgings by opening doors in the bottom of the hull.

Ice Age – Usually refers to the four or more times during the Pleistocene Epoch when the large continental glaciers formed and spread over the polar parts of the continents. Several pre-Pleistocene Ice Ages are also known.

igneous – All rocks that at one time were molten. Includes lavas (volcanic) rocks that erupt on the land surface as well as those plutonic rocks that congeal at depth.

joint – A crack in any rock along which there has been no offset. May occur in sets with similar orientations.

Kieselguhr – The German name for siliceous earth or diatomite.

knocker – A large, discrete block of rock surrounded by rocks of different nature. Common in **melanges** as seen in the Franciscan Formation.

laminated – Very thin, almost paper-like layering seen in some sedimentary rocks such as those of the Monterey and Sisquoc formations.

laterite – A brick-red soil composed mostly of red iron oxides. A common product of tropical weathering in a seasonally wet-dry climate.

left slip – The direction of displacement along a strike-slip fault like the Santa Ynez in which an observer, looking across the fault, would see evidence that displacement was to the left.

limonite – A yellowish-brown hydrous iron oxide. It is very common and its composition is very similar to iron rust. One of the most ubiquitous coloring agents in rocks. It gives rise to tan, yellow, buff or ochre colors. An important mineral pigment.

lithification – Any process that converts loose sediment into rock. Commonly due to cementation by lime (calcium carbonate) iron oxides, or to compaction by burial.

long-period wave – In the sea, waves from distant sources. Typically with 15-25 second intervals between the passage of wave crests by a fixed point. On the coast of Santa Barbara County, most of these long-period waves originate in the southern hemisphere during their winter and reach our shores during northern summer.

meanders – Sinuous curves in a stream.

melange– A heterogeneous mixture of rock types, often sheared or crushed during the subduction process, mostly fine-grained, but commonly with large, intact blocks of rock (**knockers**). Very common in the Franciscan Formation.

metamorphism – The process by which any rock can be crystallized or re-crystallized without melting. Due to any combination of heat, pressure or chemical alteration. If melting occurs, the rock, by definition, is igneous.

metavolcanic – A group of dense, fine-grained, durable rocks produced by metamorphosis of volcanic rock like basalt.

methane clathrate – See **gas hydrate**.

normal fault – An inclined fault along which the overlying (hanging wall) block appears to have moved down with respect to the underlying (foot wall) block.

older alluvium – Former stream deposits, now out of reach of modern streams. Often slightly consolidated.

olivine – An iron, magnesium silicate mineral that is green in color. Common in some basalts. The gem variety is called peridot.

ooze – An unconsolidated marine sediment containing at least 30% organic material.

ophiolite – A sequence of igneous and sedimentary materials that form on the deep sea floor. Land exposures are always related to subduction. Typically include basalt, gabbro, peridotite and deep-water bedded chert.

oreodont – A pig-like mammal that roamed North America in large herds from Late Eocene to Pliocene time.

orogeny – Mountain building or the uplift and formation of mountains.

overturn – Intense folding, usually of sedimentary rocks in which some beds have been tilted beyond the perpendicular.

paleomagnetism – Fossil magnetism. Refers to remnant magnetism of minerals in rocks that can be used to deduce changes in the earth's magnetic field or changes in the position of continents or smaller areas with respect to one another.

peridotite – An iron and magnesium-rich rock composed chiefly of the mineral olivine. Also termed an ultra-basic rock, i.e. one with a low silica content. See also **dunite**.

pillow structure – A structure that resembles a stack of pillows, formed when lava is extruded under water. Most examples are basalts erupted under the sea. Also called **pillow basalt.**

pinch-out – A characteristic of sedimentary rock in which a given bed or layer decreases in thickness, eventually disappearing.

pitch – Another name for tar in tar seeps. See also **asphaltum**.

phytoplankton – Floating plants in the sea, chiefly diatoms that are sometimes described as the "grass of the sea". The basis of the marine food chain.

plate – A unit of the outer crust of the earth, often of continental size.

plate tectonics – Refers to the interactions of crustal plates, including subduction, sea-floor spreading, transform faulting and many types of mountain building.

plug – A columnar mass of rock usually volcanic that forms the core of a volcanic mountain or an intrusion into other rocks. The Devils Tower in Wyoming is a spectacular example.

plutonic – Igneous rocks that are emplaced and congealed beneath the surface in contrast to volcanic rocks that form on or near the surface of the earth.

pozzolan – A volcanic ash-containing cement that will harden under water.

pug – See gouge.

pull-apart – Sags or depressions formed in fault zones where the crust between offset strike-slip faults is stretched (Figure 53). See also **sag pond**.

pumice – A bubbly, glassy volcanic rock most often formed from silica-rich lavas. Pumice fragments typically float on water.

quartz – Crystalline silica (SiO_2). The most common mineral species in crustal rocks. Hard, durable and chemically stable. Many varieties.

quartzite – Either a sedimentary, silica-cemented sandstone, or a silica-rich metamorphic rock derived from sandstone (metaquartzite). Both rock types are hard and durable.

radiolarians – Microscopic zooplankton that secrete siliceous tests (shells). May dominate some deep-water marine sediments where conditions favor solution of the more common calcareous organisms. Often an important source of chert, as in the Franciscan Formation.

radon – An inert, radioactive gas, a daughter element of radium. A part of the thorium series of radioactive elements. It has a half-life of 3.8 days. Compare with source bed.

reservoir rock – Porous and permeable rocks in which petroleum and natural gas accumulate. Compare with **source bed**.

reverse fault – An inclined fault in which one block has ridden up and over the adjacent block. This is a response to crustal shortening or lateral compression. If the angle of the fault plane is less than about 20°, it is called a **thrust fault**.

rhyolite – A volcanic rock rich in silica; the equivalent of plutonic granite. Usually light-colored and often associated with pumice and obsidian (volcanic glass).

right slip – The direction of displacement on a strike-slip fault like the San Andreas in which an observer, looking across the fault would see evidence that the displacement was to the right.

ripple marks – Evenly spaced small ridges produced in loose sediment by wind or water moving over the surface.

rip-rap – A ridge of loose, durable rock fragments placed along stream channels or beaches to protect various structures from erosion.

sag pond – A pull-apart structure common where strike-slip faults are offset in an echelon pattern, and where the land surface is depressed below the water table at least part of the year. See also **pull-apart**.

Salinian Block – A slice of crust in the southern Coast Ranges of California with granitic basement. It lies between the San Andreas fault on the east and the Sur-Nacimiento-Rinconada fault system on the west. Thought by some geologists to be the link between the Sierra Nevada and the Peninsular Ranges.

saltation – A jumping or bouncing path followed by blowing sand.

salt marsh – Usually a coastal marshy area connected to the sea by a system of channels.

salt weathering – A destructive process in which the sea spray from the ocean soaks into surficial rocks. As the water evaporates, the salt crystals grow and slowly wedge the rock grains apart and cause the rock to gradually crumble.

sand dollar – A disc-shaped marine animal with a calcareous shell. Related to sea urchins and star fishes.

sand spit – A sandy ridge or bar deposited by the longshore current. It often forms across the mouths of shallow bays.

sandstone – A common sedimentary rock composed of sand grains cemented together.

sea floor spreading – The process by which new oceanic crust is formed at a sea floor structure. Basaltic lava wells up from depth and forces the crustal blocks to move away from each other.

sea wall – Any sort of protective barrier installed to protect coastal areas and structures from wave erosion.

schist – A crudely layered crystalline metamorphic rock rich in platy minerals such as the micas. May be derived either from sedimentary or igneous predecessors.

scoria – A dark-colored, bubbly volcanic rock, most often basalt. Unlike pumice, scoria does not float.

sedimentary – One of the three main classes of rocks. Includes all rocks derived from incoherent materials and deposited by water, wind or ice.

seeps – Places where tar, petroleum, natural gas or water ooze out of the ground. Similar to springs.

serpentine – A soft, shiny magnesium silicate mineral often derived from metamorphism of basalt. Where the mineral occurs in large bodies it is given the rock name **serpentinite**. Usually greenish or black.

shale – A distinctly layered, very fine-grained sedimentary rock derived from muds or silts. Where layering is not prominent, the rock is called a **mudstone**.

short period waves – In the sea, usually waves with a local origin. Typically with 6-9 second intervals between passage of successive crests by a fixed point.

sill – A volcanic rock injected between and parallel to enclosing sedimentary layers. Sills are sheet-like or tabular. Compare with **dike**.

slip face – The lee slope of an active sand dune. In dry sand the inclination of 33-34° is always useful in deducing direction of sand-driving winds.

solfatera – A vent, often volcanic, from which gases and vapors are emitted. Also called a **fumarole**.

source bed – A sedimentary rock, rich in organic material, from which petroleum and natural gas are derived. Compare with **reservoir rock**.

spheroidal weathering – A form of chemical weathering in which concentric or spherical shells are produced. Commonest in granitic rocks, but may also develop in sedimentary shales or mudstones.

spit – See **sand spit**.

spreading center – The junction of two crustal plates, usually on the sea floor where new crustal material, typically basalt, wells up from the mantle, shoving the plates apart.

spring sapping – An erosional process produced by the discharge of springs, usually on a cliff face.

stratigraphic trap – Often produced by an impermeable rock capping the upturned ends of an inclined reservoir rock. May also result from the pinching out of a reservoir rock between two impermeable beds. Although these are often important sources of petroleum and natural gas, they are not evident at the surface.

Glossary

strike-slip fault – A fault along which the dominant movement is horizontal. A special category, where such a fault offsets a spreading center, is called a **transform fault**. The San Andreas fault is both a transform fault <u>and</u> a right-slip fault. See also **right-slip** and **left-slip** faults.

subduction – The process by which one of the large crustal plates is dragged or forced beneath an adjacent plate. The overlying plate is deformed into mountains in which volcanic and plutonic igneous activity is prominent.

suction dredge – A dredging machine that sucks up dredged material and pumps it to a disposal area, often a beach.

suspension – Refers to wind or air currents whose velocity is sufficient to keep sand, etc. suspended in the medium.

syncline – A downward fold, basin or canoe-shaped in which rocks on the sides dip downward toward the center of the fold. Compare with **anticline**.

tar seep – A place where petroleum or asphaltum is extruded onto the surface from reservoir rocks at depth. La Brea Tar Pits in Los Angeles is a well-known example. Many early oil fields were discovered by drilling near seeps.

tectonics – Refers to all the processes and resulting structures that are associated with plate interactions, faulting or folding.

test – A shell-like structure secreted by microscopic animals such as foraminiferans and radiolarians. Most often composed of calcium carbonate or silica.

thrust fault – A low-angled fracture along which one crustal block has ridden up and over the adjacent block. Compare with **reverse fault**.

tidal wave – See tsunami.

tombolo – A coastal sand spit connecting the shore with an offshore island or reef.

tonalite – A plutonic igneous rock similar to granite, but with less silica and richer in the dark minerals biotite and hornblende.

transform fault – A fault of dominantly horizontal displacement that offsets a **spreading center**, usually at right angles to the trend of the center. Compare with **strike-slip fault**.

travertine – A cool-water deposit of calcium carbonate. In many cases the rock is spongy or porous and is formed by the activity of lime-secreting algae.

tsumani – Very long period waves (10 to 20 minutes) created in the sea by volcanic activity, submarine landslides and by sea-floor faulting. Velocities range from 250 to 500 miles per hour. Tsunamis may create major damage along coastlines. Often called a **tidal wave**.

tube worm – One of the spectacular organisms discovered clustering around **black smokers** on the ocean floor. Tube worms and associated organisms in such settings obtain their food and energy from the waters around these vents, and do not depend upon solar energy.

tuff – A fine grained sedimentary or volcaniclastic rock formed from volcanic ash. Usually rich in silica.

type locality – The geographic locality from which a formation or a new animal or plant species was first formally described.

unconformity – A break in the sedimentary record during which an erosional interval occurred. The erosional interval can be either brief or long.

upwelling – A phenomenon in the sea in which winds, frequently on western coasts, drive surface water away from shore, causing colder, deeper, nutrient-rich water to rise to the surface. Upwelling areas typically are biologically very productive.

volcanic ash – Despite its name, combustion is not involved. Ash is fine-grained eruptive material produced entirely from volcanoes.

volcaniclastic – Any sedimentary material derived entirely from volcanic materials.

warping – Gentle folding in which the beds dip less than about 20°.

water gap – A deep channel across a mountain barrier, through which a stream flows.

wild-catting – Exploration for oil or gas with little or no geological information.

wind gap – A deep channel across a mountain carrier that is no longer occupied by a stream, although it was originally formed by stream erosion. Compare with **water gap**.

window – An eroded area that displays rocks concealed beneath the rocks that surrounded the window.

Era	Period	Epoch	Approximate Age in Years at beginning of interval
CENOZOIC	QUATERNARY	Holocene (Recent)	10,000
		Pleistocene	1.8 million
	TERTIARY	Pliocene	5 million
		Miocene	23.5 million
		Oligocene	39 million
		Eocene	53.5 million
		Paleocene	65 million
MESOZOIC	CRETACEOUS		144 million
	JURASSIC		208 million
	TRIASSIC		245 million

Table 1. Geological time scale. No rocks in Santa Barbara County are older than Jurassic in age.

REFERENCES

As noted in the preface, there is now extensive published literature dealing with many aspects of Santa Barbara County geology. Part of this literature is an array of excellent geological maps, notably those published by the Dibblee Geological Foundation. The Dibblee maps cover fully the western and southern parts of the county and several of the islands. Nearly all the northeastern part of the county is covered by maps published by the U. S. Geological Survey. In addition, the California Division of Mines and Geology has published small-scale maps covering the entire county. These are the Los Angeles, Long Beach, Santa Maria, and San Luis Obispo 1: 250,000 sheets.

The list of references that follows is an abbreviated and selected one in which the items are either reports that cover fairly large parts of the county, deal with matters of general interest, or are articles directed toward the non-specialist.

For those seeking more detailed information, many pertinent additional references will be found in the items cited below.

Atwater, Tanya
    1970   Implications of plate tectonics for the cenozoic evolution of Western North America. Geological Society of America Bulletin 81:3513-3536.

Bramlette, M. N.
    1946   Monterey formation of California and origin of its siliceous rocks. U. S. Geological Survey Professional Paper, no. 212. Washington, D.C.: G.P.O. 57 pp.

Burnett, John L.
    1991   Diatoms--the forage of the sea. California Geology 44(4):75-81.

Churchill, Ron
    1997   Radon mapping, Santa Barbara and Ventura Counties. California Geology 50(6):167-177.

Dibblee, Thomas W., Jr.
    1950   Geology of the southwestern Santa Barbara County, California. California Division of Mines Bulletin 150. 95 pp.

Dibblee, Thomas W., Jr.
    1966   Geology of the central Santa Ynez Mountains, Santa Barbara County, California. California Division of Mines and Geology Bulletin 186. 99 pp.

Dickinson, William R., Hopson, Clifford A., and Saleeby, Jason B.
    1996   Alternate origins of the coast range ophiolite (California): introduction and implications. GSA Today 6:1-10.

Ernst, W. G., ed.
    1981   The geotectonic development of California, Englewood Cliffs, N. J.: Prentice Hall. 706 pp.

Fife, Donald L. and John A. Minch, eds.
    1982   Geology and mineral wealth of the California transverse ranges. Santa Ana, Calif.: South Coast Geological Society. 699 pp.

Hill, Mary
    1999   Gold, the California story. Berkeley, Calif: University of California Press. 306 pp.

Jahns, Richard H., ed.
    1954   Geology of Southern California. California Division of Mines Bulletin 170. 2 vol.

Jenkins, Olaf Pitt, ed.
    1943   Geologic formations and economic development of the oil and gas fields of California. California Division of Mines Bulletin 118. 773 pp.

Johnson, D. L.
    1967   Caliche on the Channel Islands. California Mineral Information Service 20:151-158.

Keller, Edward A., L. D. Gurrola, J. G. Metcalf, and T. W. Dibblee, Jr.
 1995 Earthquake hazard of the Santa Barbara fold belt, California. University California, Santa Barbara Field Guide. 58 pp.

Kunitomi, Dale S., Thomas E. Hopps, and James M. Galloway
 1998 Structure and petroleum geology, Santa Barbara Channel, California. American Association of Petroleum Geologists. Pacific Section. Miscellaneous Publication 46: 328 pp.

Lewis, Lavon, P. Hubbar, E. G. Heath, and A. Pace, eds.
 1991 Southern coast ranges. South Coast Geological Society Annual Field Trip Guide Book 19: 478 pp.

Luyendyk, Bruce C., M. J. Kammerling, and R. R. Terres
 1980 Geometric model for neogene crustal rotations in Southern California. Geological Society of America Bulletin 91:211-217.

Norris, Robert M., and Robert W. Webb
 1990 Geology of California. New York: John Wiley & Sons. 541 pp.

Norris, Robert M.
 1990 Sea cliff erosion: a major dilemma. California Geology 43(8): 171-177.

Norris, Robert M.
 1991 A visit to Santa Barbara Island. California Geology 44(7): 147-151.

Olsen, Phil G., and Arthur G. Sylvester
 1975 The Santa Barbara earthquake of 29 June 1925. California Geology 28(6): 123-131.

Power, Dennis M., ed.
 1980 The California Islands: proceedings of a multidisciplinary symposium. Santa Barbara, Calif.: Santa Barbara Museum of Natural History. 787 pp.

Priestaff, Iris
    1979   Natural tar seeps and asphalt deposits of Santa Barbara County. California Geology 32(8):163-169.

Sharp, Robert P.
    1978   Coastal Southern California. Dubuque, Iowa: Kendall/Hunt, 268 pp.

Sharp, Robert P., and Allen F. Glazner
    1993   Geology underfoot in Southern California. Missoula, Mont.: Mountain Press. 224 pp.

Sylvester, Arthur G., and G. C. Brown, eds.
    1988   Santa Barbara and Ventura basins. Coast Geology Society Guidebook 64. 166 pp.

Vedder, J. G., H. C. Wagner, and J. E. Schoelhammer.
    1969   Geologic framework of the Santa Barbara Channel region. U. S. Geological Survey Professional Paper 679A : 1-11.

Weaver, Donald W.
    1969   Geology of the Northern Channel Islands. Special Publication, Pacific Section, American Association of Petroleum Geologists and SEPM. 200 pp.

Weigand, Peter W., ed.
    1998   Contributions to the geology of the Northern Channel Islands, Southern California. Bakersfield, Calif.: Pacific Section, American Association of Petroleum Geologists. MP-45: 196 pp.

Woodring, W. P., and M. N. Bramlette
    1950   Geology and paleontology of the Santa Maria district, California. U. S. Geological Survey Professional Paper 222. 185 pp.

INDEX

A

Acachuma Mine 126
Agua Caliente Canyon 80
air pollution 138
Airox Mine 125, 131, 190
Alamo Pintado Creek 22, 169, 178, 185
Alamos tonalite 63
Alaska 2, 4
Alcatraz Asphaltum Company 131
Alegria Canyon 102
Alegria Formation 65, 82, 88, 163
Alisal Road 142
Anacapa Island 5, 117
Andree Clark Refuge 153
Angostura Pass 174
Anita Formation 65, 78, 142
Antarctica 82, 125
antecedent stream 28
anticlines 15
   Brush Peak 160
   Flores 20
   Jalama 178
   Pacifico 177
Arguello 176
Arroyo Burro Creek 27, 29, 32, 153
Arroyo Burro trail 175
Arroyo Burro water gap 153
Arroyo Hondo Bridge 156
Arroyo Parida fault 100, 151, 152
arsenic 143
Arvin 111
asphaltum 128
asteroid 76
Atascadero Creek 54, 94, 100, 154, 166

B

badlands 25, 155, 165, 192, 199
   Cuyama badlands 206
Bakersfield 111
Ballard Canyon 186
Ballard Creek 187
Ballinger Canyon Road 206
Barham Ranch Oil Field 157
barite mine
   White Elephant 127, 195
Barka Slough 130, 165, 190, 191
basalt 204
basement 73
Bates Canyon 77
Bath House Beach 50, 95
Bay Point 120
beaches 33
   Bath House Beach 50, 95
   Carpinteria State Beach 129
   East Beach 38
   El Capitan State Beach 55, 56, 155
   Gaviota State Beach 42, 142
   Leadbetter Beach 40, 41, 50
   More Mesa Beach 51, 52, 128
   Pismo Beach 34, 149
   Point Sal Beach 201
   Refugio Beach 55
   Summerland Beach 49, 95
beach drift 35
beach sand cell 36
Bee Rock 128
Beechers Bay Formation 64, 118
Big Bend 5
Big Caliente Creek 172
Big Pine fault 86, 97, 104, 106
Big Pine Mountain 14, 24, 76, 104, 168, 175

Black Mountain 119
black sands 126
black smoker 71, 72, 144
Blanca Formation 63, 116, 117
Blue Canyon 72, 77, 78, 102, 170, 172
Blue Canyon Pass 70, 79, 171
blue glaucophane schist 70
Blue Oaks 145
Bluff Camp 104
Botanic Garden 96, 160
boulder-pile weathering 160
Branch Canyon 26, 107
Broadcast Peak 13, 80, 160, 161, 175
Brush Peak anticline 160
Buckhorn Formation 199
Buellton 23, 142, 157, 166, 169, 178, 197
burning tar deposit 130
Burton Mesa 164

C

Cachuma Canyon 104
Cachuma Creek 184
Cachuma fault 184
Cachuma Forest Service Campground 184
Cachuma Formation 183, 195, 196, 200
Cachuma Lake 21, 23, 31, 103, 141, 157, 168, 180
Cachuma Lake County Park 128, 169
Cachuma Mountain 76, 126
Cachuma Saddle 126, 184
Caliente Range 87, 205, 206
California Coast ranges 13, 30, 104 (See Coast Ranges)
Camelback Hill 20
Camino Cielo Road 86, 161, 170, 172, 174
campgrounds
    Cachuma Forest Service 184
    Colson Forest Service 195
    Dutch Oven Forest Service 106
    Figueroa Forest Service 184
    Fremont Forest Service 180
    Juncal Forest Service 80, 172
    Lazy Forest Service 195
    Mono Forest Service 170, 172
    P-Bar Forest Service 172
    Upper Oso Forest Service 180, 181
    Wagon Flat Forest Service 195
Campus Lagoon 53
Campus Point (see Goleta Point)
Camuesa fault 72, 104, 105, 168, 175, 183, 184
Cañada de las Calaveras 189
Cañada del Capitan Creek 55, 155
Cañada del Medio 21, 114
Cañada del Puerto 114
canyons
    Agua Caliente Canyon 80
    Alegria Canyon 102
    Arroyo Burro Canyon 32
    Ballard Canyon 186
    Bates Canyon 77
    Blue Canyon 72, 77, 78, 102, 170, 172
    Branch Canyon 26, 107
    Cachuma Canyon 104
    Colson Canyon 15, 127, 194, 200
    Corralitos Canyon 20, 165, 201
    Drum Canyon 19, 179, 189
    Foxen Canyon 103, 169, 186
    Glen Annie Canyon 85, 154
    Graciosa Canyon 165, 193
    Happy Canyon 182
    Harris Canyon 193
    Hot Springs Canyon 32
    Howard Canyon 189
    Hueneme Submarine Canyon 36
    Jalama Canyon 78, 102
    La Brea Canyon 15, 127, 132, 194, 195, 200
    Los Laureles Canyon 167
    Manzana Canyon 74
    Mine Canyon 106
    Mission Canyon 160
    Morales Canyon 205

Newsome Canyon 127
Nojoqui Canyon 78
Purisima Canyon 194
Quatal Canyon 207
Rattlesnake Canyon 84, 173
Rhoda Canyon 107
San Lucia Canyon 164
San Pedro Canyon 160
Santa Barbara Canyon 107, 206
Shuman Canyon 191
Suey Canyon 200
Sycamore Canyon 84, 96, 173, 185
Tepusquet Canyon 16, 127, 199
Toro Canyon 152
Vaqueros Canyon 26
Cape Mendocino 2
Careaga Formation 19, 21, 66, 93, 131, 157, 165, 168, 182, 186, 187, 189, 191, 192, 193
Carneros Creek 54
Carpinteria 33, 38, 49, 93, 107, 108, 175
Carpinteria fault 48, 100, 101, 150
Carpinteria plain 152
Carpinteria Salt Marsh 48, 49, 151
Carpinteria State Beach 129
Carpinteria tar seep 131
Carpinteria Thunderbowl 48, 101, 151
Carpinteria Valley 33
Carrillo Hotel 111
Cascade Range 91
Casitas Creek 95
Casitas Formation 49, 50, 95, 107, 152
Casmalia 24, 93, 134, 190, 191
Casmalia Hills 20, 44, 165
Casmalia Oil Field 20, 190, 191
Castle Rock 120
Cat Canyon Oil Field 21, 134, 157, 188
Catway Road 184
Celite Quarry 124, 169, 198
Central Valley 4, 22, 99
Central Valley of Santa Cruz Island (Cañada del Medio) 21, 22, 114

Cerro Noroeste 206
Channel Islands 9, 12, 97, 113, 120
Channel Islands National Park 113
chemical sediments 61
chert
  radiolarian 70, 184
Chinese Harbor 115, 130
Christi Ranch 114, 115
chromite 72, 126, 182
Chumash 128
Cieneguitas Creek 27, 100, 166
cinnabar 126
Circle Bar-B Guest Ranch 159
clastic 61
Clear Creek 106, 196
Clear Lake 14
coal deposit 130
Coal Oil Point 43, 51, 54, 55, 135, 138
Coast Range ophiolite 127, 194
Coast Range orogeny 93
Coast Ranges 4, 5, 8, 69, 74, 86, 97, 103, 200, 202 (See California Coast Ranges)
coastal erosion 39
coastal plain 33, 93
cold seep 128
Cold Spring Arch bridge 167
Cold Springs Saddle 171
Coldwater Formation 13, 56, 80, 84, 96, 108, 135, 160, 161, 167, 173, 175
Colorado Desert 4
Colson Canyon 15, 127, 194, 200
Colson Forest Service Campground 195
columnar sections 62
continental drift 1, 67
continental shelf 4
Contra Costa County 200
copper 127, 143, 182
coral reef 78
Corralitos Canyon 20, 165, 201
Corralitos Ranch 201
counties
  Contra Costa County 200
  Kern County 104, 134

Ventura County 131
Los Angeles County 84, 99, 114, 134
Monterey County 105
Orange County 120
San Benito County 126
San Bernardino County 5
San Diego County 90, 116, 120
Santa Barbara County 39, 100, 109
Santa Clara County 126
Ventura County 9, 24, 25, 34, 44, 84, 85, 86, 99, 100, 102, 104, 105, 106, 108, 114, 119, 120, 127, 131, 134, 136, 142, 151, 173, 206
county parks
　Cachuma Lake County Park 128, 169
　Miguelito County Park 170
　Nojoqui Falls 186, 188
Cozy Dell Formation 10, 14, 63, 64, 65, 80, 108, 118, 159, 161, 162, 170, 173, 175, 177, 180
creeks
　Alamo Pintado Creek 22, 169, 178, 185
　Arroyo Burro Creek 27, 29, 153
　Atascadero Creek 54, 94
　Ballard Creek 187
　Big Caliente Creek 172
　Cachuma Creek 184
　Carneros Creek 54
　Casitas Creek 95
　Cieneguitas Creek 27, 100, 166
　Clear Creek 106, 196
　Devereux Creek 54
　Dos Pueblos Creek 154
　El Jaro Creek 18, 74, 163
　Foxen Canyon Creek 187
　Foxen Creek 24
　Franklin Creek 49, 152
　Gaviota Creek 26, 142, 156
　Glen Annie Creek 154
　Harris Creek 165
　Honda Creek 72, 74
　Hot Springs Creek 143
　Indian Creek 78, 106
　Jalama Creek 57, 77, 176
　La Hoya Creek 177
　Los Carneros Creek 154
　Manzana Creek 76
　Maria Ygnacio Creek 54, 154
　Mission Creek 28, 144, 153
　Mono Creek 78
　Nojoqui Creek 142, 188
　Orcutt Creek 165, 193
　Picay Creek 100
　Pozo Creek 116
　Quilota Creek 158
　Refugio Creek 55, 155, 158
　Rincon Creek 47, 48, 95, 100, 142, 149
　Romero Creek 50, 153
　Salsipuedes Creek 18, 77, 164, 177, 198
　San Antonio Creek 24, 34, 58, 157, 165, 190, 191, 193
　San Jose Creek 54, 154
　San Pedro Creek 54, 154
　San Roque Creek 28
　San Ysidro Creek 50
　Santa Agueda Creek 22, 169, 182
　Santa Monica Creek 49, 151
　Santa Rosa Creek 190
　Sespe Creek 84
　Sweetwater Creek 128
　Sycamore Creek 28
　Tajiguas Creek 155, 158
　Tecololito Creek 54
　Zaca Creek 30, 72, 157, 187
　Zanja de Cota Creek 22, 178
Cuddy Valley 104
cuspate delta 48
Cuyama 25, 127, 206
Cuyama badlands 206
Cuyama Peak 207
Cuyama Phosphate Mine 127
Cuyama River 24, 25, 32, 61, 72, 106, 107, 202, 204
Cuyama River Road 199
Cuyama River Valley 8, 16, 24, 26, 87, 97, 106, 194, 196, 202, 206
Cuyler Harbor 44, 46, 120, 121

Index

D

dacite 120
dams 21, 23
    Bradbury Dam 23, 31, 32, 141
    Gibraltar Dam 23, 141
    Juncal Dam 22, 31, 72, 141
    Twitchell Dam 19, 25
Del Mar 120
Devereux Creek 54
Devereux Slough 53, 54
Devil's Tower 120
diabase 184, 194, 200, 201, 204
diapir 200
diatomite 61, 91, 123, 164
    Celite diatomite quarry 169
    Sisquoc diatomite 163
diatomaceous shale
    Sisquoc 199, 201
diatoms 91, 123, 124
dikes 119
dimension stone 127
diorite
    Willows 63
Divide Peak 175
Divide Peak Junction 171
dolomite bed 194
Dos Cuadras Oil Field 137
Dos Pueblos Creek 154
Doulton Tunnel 31
Drum Canyon 19, 179, 189
dunites 182
Dutch Oven Campground 106

E

Eagle Mine 126
early man 119
earthquakes 4, 98, 109
    Fort Tejon, 1857 109
    Goleta, 1968 111
    Kern County, 1952 111
    Los Alamos, 1902 109
    Point Arguello, 1927 111
    San Francisco, 1906 111
    Santa Barbara, 1925 102, 110
    Santa Barbara, 1926 111
    Santa Barbara, 1941 111
    Santa Barbara, 1978 111
    Santa Barbara Channel, 1812 109
East Beach 38
East Cat Canyon Oil Field 21
East Huasna fault 195, 203
East Pacific Rise 1, 143
East Point 118
El Capitan Oil Field 155
El Capitan Point 55
El Capitan State Beach 55, 56, 155
El Estero (see Carpinteria Salt Marsh)
El Jaro Creek 18, 74, 163
El Montañon 115
elevated marine terraces 33, 75, 94
Ellwood 101, 131, 154
Elwood Oil Field 132, 135
Espada Formation 65, 66, 74, 75, 157, 163, 168, 172, 177, 181, 183, 184
expanded shale 126

F

fanglomerate 96, 166, 173
faults
    Arroyo Parida 100, 151, 152
    Big Pine 86, 104, 106
    Cachuma 184
    Camuesa 72, 104, 105, 168, 175, 183, 184
    Carpinteria 48, 100, 101, 150
    East Huasna 195, 203
    Garlock 104
    Honda 72, 170
    Lavigia 16, 17, 28, 29, 32
    Little Pine 72, 93, 103, 105, 126, 168, 174, 181, 183, 185
    Mesa 16, 17, 50, 100, 102, 109, 153
    Mission Ridge 16, 144, 153
    Morales 205, 206
    More Ranch 16, 17, 27, 51, 94, 100, 153
    Nacimiento 106
    Pacifico 102, 176
    Red Mountain 100
    Refugio 159

Rinconada 106, 196
Rincon Creek-Carpinteria 48
San Andreas 2, 5, 67, 73, 97, 98, 104, 106
Santa Cruz Island 21, 99, 114, 115, 118
Santa Rosa Island 99, 118
Santa Ynez 9, 97, 102, 143, 156, 159, 163, 166, 168, 171, 176
Santa Ynez River fault 103
South Cuyama 206
strike-slip faults 98
Sur-Nacimiento 73, 106, 195, 204
thrust fault 98
transform 2
Fernald Point 50, 153
Figueroa Forest Service Campground 184
Figueroa Mountain 7, 15, 60, 61, 62, 71, 72, 105, 126, 143, 145, 168, 182
Five Points Shopping Center 100, 101
Flandrian Transgression 26
Flores anticline 20
Flores Flat 170, 171, 174
folds 107
Forbush Flat 77, 78, 102, 171
Ford Point 118
formations 67
  Alegria Formation 65, 82, 88, 163
  Anita Formation 65, 78, 142
  Beechers Bay Formation 118
  Blanca Formation 63, 116, 117
  Buckhorn Formation 199
  Cachuma Formation 183, 195, 196, 199
  Careaga Formation 19, 21, 66, 157, 168, 193
  Casitas Formation 49, 50, 95, 107, 152
  Coldwater Formation 56, 80, 84, 108, 167
  Cozy Dell Formation 10, 14, 63, 64, 65, 80, 108, 118, 161, 175

Espada Formation 65, 66, 74, 75, 163, 172, 181, 183, 184
Foxen Formation 66
Franciscan Formation 8, 65, 66, 74, 86, 117, 143, 171, 175, 182, 204
Gaviota Formation 65, 88, 142, 155, 156
Gaviota-Sacate Formation 170, 197
Honda Formation 65, 74, 7
Jalama Formation 65, 74, 142, 171, 177
Jolla Vieja Formation 63, 116
Juncal Formation 80, 171, 17
Knoxville Formation (See Espada Formation)
Lospe Formation 20, 66, 82, 201
Matilija Formation 56, 65, 80, 155, 162, 170, 174, 177, 181
Monterey Formation 16, 19, 24, 31, 32, 43, 44, 48, 55, 57, 58, 60, 63, 64, 65, 66, 67, 68, 90, 96, 105, 115, 117, 119, 124, 125, 127, 128, 130, 131, 135, 145, 150, 154, 155, 156, 158, 159, 165, 170, 176, 178, 179, 180, 185, 186, 194, 195, 198, 199, 202, 206
Morales Formation 17, 206
Morris Formations 199
Orcutt Foundation 66
Paso Robles Formation 19, 21, 24, 66, 95, 105, 157, 165, 169, 179, 182, 185, 187, 188, 189, 192, 193
Point Sal Formation 201
Poway Formation 116, 120
Pozo Formation 63, 77
Rincon Formation 29, 50, 55, 63, 64, 65, 88, 89, 135, 145, 156, 18
Sacate Formation 13, 65, 81, 155, 159, 161, 162
Santa Barbara Formation 50, 52, 94, 101, 107, 115, 153, 154
Santa Margarita Formation 127, 205

Index

Sespe Formation 64, 65, 118, 152, 153, 156, 159, 163, 166, 175, 173, 180, 197, 200, 20
Sierra Blanca Formation 65, 78, 79, 116
Simmler Formation 87, 195, 199, 204, 206
Sisquoc Formation 18, 51, 65, 66, 90, 123, 124, 125, 130, 150, 165, 168, 178, 179, 182, 187, 189, 191, 193, 194, 198
South Point Formation 64
Temblor Formation 75, 172
Tranquillon Formation 58, 65, 178, 185, 188, 197, 201
Vaqueros Formation 63, 64, 65, 87, 88, 135, 155, 156, 163, 180, 188, 195, 197, 200, 204
fossil oysters 84, 173
Fox Mountain 107, 206
Foxen Canyon Creek 24, 103, 169, 186, 187
Foxen Formation 66, 190
Foxen mudstone 191
Franciscan Formation 8, 9, 65, 66, 72, 74, 82, 86, 103, 117, 143, 168, 171, 175, 182, 184, 203, 204
Franklin Creek 50, 152
Fremont Forest Service Campground 180
Fugler Point 24, 186

G

gabbro 70
Garey 24, 127, 186, 200
Garlock fault 104
gas fields
  La Goleta Gas Field 138
gas hydrates 139
gas seep 54
Gato Ridge 20
Gaviota 32, 78, 82, 89, 156
Gaviota Hot Springs 156
Gaviota Creek 8, 26, 142, 156
Gaviota Formation 27, 65, 88, 142, 155, 156, 169

Gaviota Pass 13, 18, 84, 85, 88, 102, 142, 155
Gaviota State Beach 42, 142, 156
Gaviota-Sacate Formation 170, 197
geologic control of vegetation 145
geothermal gradient 141
ghost tree 44, 47, 121
Gibraltar Reservoir 9, 23, 31, 61, 72, 74, 103, 126, 174, 180
Gibraltar Road 82, 142, 172
glaucophane schist 70, 116
Glen Annie Canyon 85, 154
Glen Annie Reservoir 31, 141
gold 126
Goleta 33, 39, 51, 52, 93, 94, 95, 132, 145
Goleta Point 51, 54, 55, 112, 138
Goleta Slough 17, 26, 53, 54, 101, 138, 154
Goleta Valley 33, 101
Government Point 57
Graciosa Canyon 165, 193
Graciosa Ridge 20
graywacke 69
Grefco Quarry 163, 198
Guadalupe 158, 162
Guadalupe Dunes 44, 127, 144, 166
Gulf of California 92, 125
gusher 134

H

Happy Canyon 182
harbor
  Cuyler Harbor 44, 46, 120, 12
  Chinese Harbor 115, 130
  Potato Harbor 115
  Prisoners Harbor 21, 114
  Santa Barbara Harbor 37, 38, 39, 41, 46, 146, 152
Harris Point 120
Harris Canyon 193
Harris Creek 165
Harris Grade 19, 164, 193, 194
Harris Point 120
Hartnell No. 1 133, 134
Hawaii 91

headlands 47
Hermenegildo Sal 60
highways
  State Highway 1 162, 190, 191, 193, 198
  State Highway 135 190, 193
  State Highway 154 166
  State Highway 166 196, 199, 202
  State Highway 246 178, 189, 198
  State Highway 33 202, 206
  U.S. Highway 101 149, 192, 197, 202
Hollister Ranch 78, 102
homocline 9
Honda Creek 72, 74
Honda fault 72, 170
Honda Formation 65, 74, 76
Honda Valley 169
Hope Ranch 17, 27, 32, 86, 93, 95, 108, 146, 153
hot springs 32
Hot Springs Canyon 32
Hot Springs Creek 143
Howard Canyon 189
Hueneme Submarine Canyon 36
Hurricane Deck 77, 195

I

ice-age 51
igneous rocks 61
Indian Creek 78, 106
iridium 76
Isla Vista 17, 54, 93, 146, 147, 148
islands
  Anacapa Island 5, 117
  Channel Islands 9, 12, 97, 113, 120
  Mescalitan Island 100, 154
  San Miguel Island 4, 5, 44, 46, 47, 99, 116, 120
  San Nicolas Island 44, 116, 120, 121
  Santa Barbara Island 113, 114
  Santa Catalina Island 117
  Santa Cruz Island 4, 5, 12, 44, 77, 86, 99, 114, 118, 120, 130
  Santa Rosa Island 4, 5, 12, 44, 64, 86, 116, 117, 118, 119
  Sutil Island 113
island block 120

J

Jalachichi Summit 77, 177
Jalama anticline 178
Jalama Beach County Park 176, 178
Jalama Canyon 78, 102
Jalama Creek 57, 77, 176
Jalama Formation 65, 74, 142, 171, 177
Jalama Ranch 177
Jameson Lake 9, 23, 31
joints 98
Jolla Vieja Formation 63, 116
Juncal Dam 22, 31, 72, 141
Juncal Forest Service Campground 80, 172
Juncal Formation 80, 171, 172, 174
Juncal shale 172, 174

K

Kern County 104, 134
kieselguhr (diotomite) 92
King City 106
Kinton Point 116
knocker 8, 183, 185, 204
Knoxville Formation 66

L

La Brea Canyon 15, 127, 132, 194, 195, 200
La Cumbre Middle School 95
La Cumbre Peak 80, 174, 175, 181
La Goleta Gas Field 138
La Hoya Creek 177
La Mesa 107
La Playa field 37
La Purisima Concepcion 109
Laguna Blanca 27, 32

Index

lakes 30
  Cachuma Lake 23, 31, 103, 141, 157, 180
  Clear Lake 14
  Gibraltar Lake 9
  Jameson Lake 9, 23, 31
  Zaca Lake 30, 31, 103, 127, 185
landslide 89, 130, 163, 170, 183, 198, 200, 205
Las Cruces 32, 143, 162
  (see Gaviota Hot Springs)
Las Positas Park 28, 95
Laurel Canyon syncline 167
Lavigia fault 16, 28, 29, 32
Lavigia Hill 107, 153
Lazy Campground 195
Leadbetter Beach 40, 41, 50
limestone
  Sierra Blanca limestone 172
lithification 94
Little Pine fault 72, 93, 103, 105, 126, 168, 174, 181, 183, 185
Little Pine Mountain 75, 167, 168, 180
Lion Rock 58
Live Oak Forest Service Picnic Area 180, 181
Lizard Head 104
Lockwood Valley 105
Loma Alta 167, 168, 180
Lompoc 22, 72, 74, 77, 92, 93, 103, 123, 134, 143, 162, 164, 169, 178, 179, 192
Lompoc Hills 18
Lompoc Narrows 22
Lompoc Oil Field 19, 194
Lompoc Valley 178
longshore current 35
Loon Point 50, 95, 107, 132, 152
Loreto Plaza 100
Los Alamos 112, 125, 131, 134, 166, 189, 190, 193
Los Alamos syncline 157, 190, 191
Los Alamos Valley 19, 20, 24, 96, 108, 193
Los Carneros Creek 154
Los Coches Mountain 16
Los Laureles Canyon 167
Los Olivos 22, 149, 169, 185, 186, 187
Los Prietos Ranger Station 180
Lospe Formation 20, 66, 82, 201

M

Madulce Peak 106, 175
Manzana Canyon 74
Manzana Creek 76
Maria Ygnacio Creek 54, 154
marine terrace 93, 118, 150, 154, 156, 178
Mariposa Reina 156
Matilija Formation 13, 56, 65, 80, 96, 135, 155, 159, 162, 170, 171, 173, 174, 175, 177, 181
Matilija overturn 142
Mattei's Tavern 169
McKinley Mountain 76, 183
McPherson Peak 106, 194
melange 70, 204
mercury 126, 182
  (see quicksilver)
mesas
  Burton Mesa 164
  More Mesa 17, 54, 94, 95, 146
  Santa Barbara Mesa 17
Mesa fault 16, 17, 50, 100, 102, 109, 153
Mesa Hills 153,
Mescalitan Island 100, 154
metal mining 126
  arsenic 143
  chromium 72, 126, 182
  copper 127, 143, 182
  gold 126
  iridium 76
  platinum 126
  quicksilver (mercury) 126, 182, 184
  zinc 127, 143
metamorphic 62
methane clathrates 139
Miguelito County Park 170

Mid-Atlantic Ridge 1
minerals
  barite 127
  chromite 126
  cinnabar 126
mines
  Acachuma Mine 126
  Airox Mine 125, 131, 190
  Cuyama Phosphate Mine 127
  Red Rock Mine 126, 184
  White Elephant barite mine 127, 195
Mine Canyon 106
mineral resources 123
Miranda Pine Forest Service Campground 196
Miranda Pine Mountain 16, 106, 194, 196
Mission Creek 28, 144, 153, 160
Mission Ridge 30, 96, 100, 144, 153
Mission Ridge fault 16, 144, 153
Mission Santa Ines 178
Mission Tunnel 31, 174
Mojave Desert 4, 25, 86, 162, 176
Mono Adobe 172
Mono Creek 78
Mono Forest Service Campground 170, 172
Montalvo trends 100
Montecito 34, 108, 141, 142
Montecito overturn 13, 98, 142, 152, 171, 172, 173
Montecito Peak 80, 171
Monterey 90, 127, 179
Monterey County 105
Monterey Formation 29, 42, 43, 44, 57, 16, 19, 24, 31, 32, 48, 55, 58, 60, 63, 64, 65, 66, 67, 68, 75, 90, 96, 105, 115, 117, 118, 119, 124, 125, 127, 128, 130, 131, 135, 143, 145, 150, 154, 155, 156, 158, 159, 163, 165, 167, 168, 170, 172, 176, 178, 180, 182, 184, 185, 186, 194, 195, 198, 199, 200, 202, 206
Morales Canyon 205
Morales fault 205, 206

Morales Formation 17, 206
Morales thrust fault 205
More Mesa 17, 54, 94, 95, 146
More Mesa Beach 51, 52, 128
More Mesa tar deposit 51
More Ranch 32, 53
More Ranch fault 16, 27, 32, 51, 53, 94, 100, 153
Morris Formation 199
Morro Bay 202
Morro Rock 120, 203
Mount Abel 206
Mount Calvary Retreat 82
Mount Diablo 200
Mount Lospe 20, 59, 165
Mount Solomon 20, 157
mountains
  Big Pine Mountain 14, 24, 76, 104, 168, 175
  Black Mountain 119
  Cachuma Mountain 76, 126
  Figueroa Mountain 7, 15, 60, 61, 62, 71, 72, 105, 126, 143, 145, 168, 182
  Fox Mountain 107, 206
  Little Pine Mountain 75, 167, 168, 180
  Los Coches Mountain 16
  McKinley Mountain 76, 183
  Miranda Pine Mountain 16
  Old Man Mountain 21
  Pine Mountain 106
  Redrock Mountain 19
  Rincon Mountain 132
  San Rafael Mountain 168
  San Rafael Mountains 13, 16, 24, 72, 74, 76, 78, 86, 93, 96, 108, 160, 161, 166, 170, 174, 181, 184, 186
  San Gabriel Mountains 99
  Santa Monica Mountains 114, 120
  Santa Ynez Mountains 5, 7, 9, 10, 13, 18, 28, 33, 36, 58, 72, 78, 79, 80, 81, 82, 85, 88, 89, 93, 95, 97, 98, 102, 104, 108, 111, 118, 121, 135, 142, 145, 150, 151, 152, 155,

Index 239

158, 166, 170, 173, 181, 186
Sierra Madre Mountains 13, 77, 87, 93, 106, 127, 194
Soledad Mountain 119
Stanley Mountain 204
Tranquillon Mountain 9, 13, 18, 58, 74, 90, 117, 178
Wildhorse Mountain 127
mudflows 96, 160, 173
mudstone
   Rincon mudstone 154, 158, 159, 177, 181, 197
Mussel Rock 144

N

Naciminento-Rinconada fault zone 73, 106
natural gas 92, 128
natural gas seeps 137
New Almaden 126
New Cuyama 25, 206
New Idria 126, 200
Newsome Canyon 127
Nipomo Dunes 44
Nojoqui Creek 78, 142, 188
Nojoqui Falls 142, 186
Nojoqui Falls County Park 186, 188
Nojoqui Summit 86, 142, 156
North American Plate 2

O

Obispo Tuff 19, 25, 185, 202
oil 128
oil field
   Barham Ranch Oil Fields 157
   Casmalia Oil Field 20, 190, 191
   Cat Canyon Oil Field 21, 134, 157, 188
   Dos Cuadras Oil Field 137
   East Cat Canyon Oil Field 21
   El Capitan Oil Field 155
   Elwood Oil Field 132, 135
   Lompoc Oil Field 19, 194
   Orcutt Oil Field 21, 133, 157
   Russell Ranch Oil Field 206

   Santa Barbara Mesa Oil Field 40, 134
   Santa Maria Valley Oil Field 135, 186, 192
   South Cuayma Oil Field 135
   Summerland Oil Field 132, 152
   West Cat Canyon Oil Field 134, 192
oil seep 54
Ojai 9, 100, 206
Ojai Valley 142
Old Man Mountain 21
Old Maud oil well 133, 134
Old San Marcos Pass Road 82
older alluvium 96, 154, 169, 178, 179, 182, 186, 187, 188
older dune sand 158, 179, 186, 193
ophiolite 20, 58, 60, 165, 201, 204
Orange County 120
Orcutt 20, 134, 193
Orcutt Creek 165, 193
Orcutt Oil Field 21, 133, 157
Orcutt sand 20, 46, 66, 144, 157, 164, 179, 186, 190, 191, 192, 194, 199, 202
organic sediments 61
Ortega Hill 50, 152
Ozena 104
Ozena fault 106, 206

P

P-Bar Forest Service Campground 172
Pacific Coast Railroad 169
Pacific Ocean 4
Pacific Plate 2
Pacifico anticline 177
Pacifico fault 102, 176
Painted Cave 167
Painted Cave syncline 167, 175
Palos Verdes Peninsula 117
Paradise Road 180
Paso Robles Formation 19, 21, 24, 66, 95, 105, 157, 165, 169, 179, 182, 185, 187, 188, 189, 192, 193
peaks
   Broadcast Peak 13, 80, 160, 161, 175
   Brush Peak 160
   Cuyama Peak 207

Divide Peak 175
La Cumbre Peak 80, 174, 175, 181
Madulce Peak 106, 175
McPherson Peak 106, 194
Montecito Peak 80, 171
Ranger Peak 184
Santa Cruz Peak 104, 183
Santa Ynez Peak 13, 80, 128, 161, 175
Zaca Peak 169, 185
peat deposit 130
Pendola Guard Station 172
Peninsular Ranges 4
peridotite 60
Peru-Chile Trench 1
phosphate deposit
  Pine Mountain
Picacho Diablo 114
Picay Creek 100
pillow basalt 59, 62, 70, 113, 120, 184, 201
Pine Flat 106, 195
Pine Mountain 106
Pismo Beach 34, 149
plant toxicity 185
plates 1
  Pacific Plate 2
platform blowout 136
platinum 126
points 47
  East Point
  Point Arguello 8, 18, 33, 34, 39, 56, 58, 88, 201
  Point Bennett 120
  Point Castillo 39, 40, 50, 95
  Point Conception 9, 18, 39, 56, 57, 81, 88, 138, 145, 176
  Point Hueneme 34, 119
  Point Pedernales 46, 58
  Point Reyes 48
  Point Sal 7, 20, 24, 43, 46, 58, 59, 61, 72, 88, 104, 127, 191, 201
  Point Sur 48
Point Sal Beach 201
Point Sal Formation 201
Point Sal Ridge 20, 59, 165, 201
Point Sal shale 201

Port San Luis (Port Harford) 169
Potato Harbor 115
Poway Formation 116, 120
Pozo Creek 116
Pozo Formation 63, 77
pozzolan 126
Precambrian granitic rock 206
Prince island 120
Prisoners Harbor 21, 114
pull-apart structure 101, 151
Punta Gorda 100, 108
Purisima Canyon 194
Purisima Hills 19, 20, 93, 108, 125, 131, 157, 194
Purisima Point 24, 43, 46, 58, 193

Q

quarries
  Celite Quarry 124, 169, 198
  Grefco Quarry 163, 198
  Tepusquet Quarry 127
quartzite 176
Quatal Canyon 207
quaternary alluvium 202
quicksilver 126, 182
  Red Rock Mine 126, 184
Quilota Creek 158

R

radiolarian chert 70, 184
radon 89
Ramajal syncline 177
Rancho San Julian 163
Rancho Tinaquaic 187
ranges
  Caliente Range 87, 205, 206
  California Coast ranges 13, 30, 104 (See Coast Ranges)
  Cascade Range 91
  Coast Ranges 4, 8, 69, 74, 86, 97, 103, 200, 202 (See California Coast Ranges)
  Peninsular Ranges 4
  Transverse Range 4, 5, 14, 67, 86, 97, 98, 103, 145
  Uinta Range 4

Index

Ranger Peak 184
Rattlesnake Canyon 84, 173
Red Mountain fault 100
Red Rock Mine 126, 184
Redrock Mountain 19
Refugio Beach 55
Refugio Creek 55, 155, 158
Refugio fault 159
Refugio Pass 13, 81, 158, 160, 162, 178
reservoirs
   Gibraltar Reservoir 23, 31, 61, 72, 74, 103, 126, 174, 180
   Glen Annie Reservoir 31, 141
   Sheffield Reservoir 31, 96, 100, 172
   Twitchell Reservoir 31, 202
reverse faults 98
Rhoda Canyon 107
Richardson Rock 120
Rincon Creek 47, 48, 95, 100, 142, 149
Rincon Creek-Carpinteria fault 48
Rincon Formation 29, 50, 55, 63, 64, 65, 88, 89, 135, 145, 154, 156, 158, 159, 177, 180, 181, 197
Rincon Mountain 132
Rincon Point 37, 39, 47, 55, 100, 108, 131, 138, 150
Rinconada fault 106, 196
rivers
   Cuyama River 24, 25, 32, 61, 72, 106, 107, 202, 204
   Santa Clara River 34
   Santa Maria River 26, 33, 44, 45, 126, 144, 158, 166, 186, 202
   Santa Ynez River 18, 25, 31, 34, 45, 46, 72, 74, 103, 126, 127, 141, 142, 157, 160, 164, 168, 169, 171, 174, 175, 179, 180, 181, 188, 197, 198
   Sisquoc River 20, 77, 93, 106, 127, 132, 158, 186, 200, 202
   Ventura River 142
river terrace 160, 198, 205
Riverside 5
Riviera 100
Riviera-Eucalyptus Hill uplift 28
roads
   Alisal Road 142
   Ballinger Canyon Road 206
   Camino Cielo Road 86, 161, 170, 172, 174
   Catway Road 184
   Cuyama River Road 199
   Gibraltar Road 82, 142, 172
   Harris Grade Road 164
   San Marcos Pass 9, 13, 81, 82, 155, 157, 160, 166, 167, 172
   Old San Marcos Pass Road 82
   Paradise Road 180
   Refugio Pass Road 158, 162
   Santa Rosa Road 197
   Sierra Madre Road 204
   Sweeney Road 198
rocks
   chert
      radiolarian 70, 184
   dacite 120
   diabase 184, 194, 200, 201, 204
   diatomite 123
   dunite 182
   fanglomerate 96, 166, 173
   graywacke 69
   limestone 172
   quartzite 176
   travertine 142, 188
Rocky Point 47, 58
Romero Creek 50, 153
Romero Saddle 171
Russell Ranch Oil Field 206

S

Sacate Formation 13, 65, 81, 155, 159, 161, 162
sag pond 100, 151
Salinian block 73, 106, 196, 204
Salisbury Potrero 16
Salsipuedes Creek 18, 77, 164, 177, 198
saltation 43
Salton Sea 2
San Andreas fault 67, 73, 97, 98,

104, 106
San Antonio Creek 20, 24, 34, 58, 157, 165, 190, 191, 193
San Benito County 126
San Bernardino County 5
San Diego 120
San Diego County 90, 116, 120
San Francisco Bay area 90
San Francisco earthquake of 18 April 1906 111
San Gabriel Mountains 99
San Jose Creek 54, 154
San Lucia Canyon 164
San Luis Obispo 25, 120
San Luis Obispo County 25, 34, 87, 106, 120, 144, 149, 158, 204
San Marcos Pass 9, 13, 81, 82, 155, 157, 160, 166, 167, 172
San Miguel Island 4, 5, 44, 46, 47, 99, 116, 120
San Miguelito syncline 164
San Nicolas Island 44, 116, 120, 121
San Onofre Breccia 63, 116, 117
San Pedro 117
San Pedro Canyon 160
San Pedro Creek 54, 154
San Rafael Mountain 168
San Rafael Mountains 13, 16, 24, 72, 74, 76, 78, 86, 93, 96, 108, 160, 161, 166, 170, 174, 181, 184, 186
San Roque 96
San Roque Creek 28
San Ysidro Creek 50
sand dunes 33, 41, 54, 178, 202
Sand Point 49, 100
sands
  black sands 126
sand spits 48
sandstone
  Careaga sandstone 131
  Coldwater sandstone 13, 96, 135, 160, 161, 173, 175
  Franciscan graywacke sandstone 203
  Gaviota sandstone 27, 156
  Hurricane Deck sandstone 200
  Matilija sandstone 13, 96, 135, 159, 171, 173, 175
  Santa Margarita sandstone 16
  Temblor sandstone 181
  Vaqueros sandstone 159
Sandpiper Golf Course 131
Sandy Point 118
Sandyland 49
Sandyland Cove 49
Santa Agueda Creek 22, 169, 182
Santa Anita 77
Santa Barbara Airport 53, 54, 111, 154
Santa Barbara Breakwater 39, 50
Santa Barbara Canyon 107, 206
Santa Barbara Channel 5, 54, 92, 97, 98, 99, 109, 128, 136, 137, 138 139, 160
Santa Barbara County 39, 100, 109
Santa Barbara Formation 50, 52, 94, 101, 107, 115, 153, 154
Santa Barbara Harbor 37, 38, 39, 41, 146, 152
Santa Barbara High School 28
Santa Barbara Island 113, 114
Santa Barbara Mesa 17
Santa Barbara Mesa Oil Field 40, 134
Santa Barbara Mission 28, 96, 128, 144
Santa Barbara Point 40, 43, 51
Santa Catalina Island 117
Santa Clara County 126
Santa Clara River 34
Santa Claus Village 38, 152
Santa Cruz Island 4, 5, 12, 44, 77, 86, 99, 114, 118, 120, 130
Santa Cruz Island fault 21, 99, 114, 115, 118
Santa Cruz Island schist 63, 69
Santa Cruz Peak 104, 183
Santa Margarita Formation 16, 127, 205
Santa Maria 15, 34, 82, 93, 127, 149, 185, 186, 190, 202
Santa Maria Basin 93
Santa Maria River 26, 33, 44, 45,

126, 144, 158, 166, 186, 202
Santa Maria Valley 20, 24, 32, 93, 132, 158, 186
Santa Maria Valley Oil Field 135, 186, 192
Santa Monica 99
Santa Monica Creek 49, 151
Santa Monica Mountains 114, 120
Santa Rita Hills 19
Santa Rita Valley 19
Santa Rosa Creek 190
Santa Rosa Hills 18, 93, 196
Santa Rosa Island 4, 5, 12, 44, 64, 86, 116, 117, 118, 119
Santa Rosa Island fault 99, 118
Santa Rosa Road 197
Santarosae 119
Santa Ynez 178
Santa Ynez fault 9, 97, 102, 143, 156, 159, 163, 166, 168, 171, 176
Santa Ynez Mountains 5, 7, 9, 10 13, 18, 28, 33, 36, 58, 72, 78, 79, 80, 81, 82, 85, 88, 89, 93, 95, 97, 98, 102, 104, 108, 111, 118, 121, 135, 142, 145, 150, 151, 152, 155, 158, 166, 170, 173, 181, 186
Santa Ynez Peak 13, 80, 128, 161, 175
Santa Ynez River 18, 24, 25, 31, 34, 45, 46, 72, 74, 96, 103, 126, 127, 141, 142, 157, 160, 164, 166, 167, 168, 169, 171, 174, 175, 179, 180, 181, 188, 196, 197, 198
Santa Ynez River fault 103
Santa Ynez River terrace 164, 180, 182, 197
Santa Ynez River Valley 8, 22, 160, 161, 170
schist
  blue glaucophane schist 70, 116
  Santa Cruz Island schist 63, 69
Sea Elephant Cove 114
sea-floor spreading 1
seacliffs 39, 146
Sedgwick Ranch Reserve 145

seep
  Carpinteria tar seep 131
  cold seep 128
  natural gas seeps 54, 137
  oil seep 54
  submarine seeps 128
  tar seep 129
serpentine 126, 145
serpentinite 62, 69, 168, 174, 183, 200, 204
Sespe Creek 84
Sespe Formation 64, 65, 118, 152, 153, 156, 159, 163, 166, 173, 175, 180, 188, 197, 200, 201
Shag Rock 113
Sheffield Reservoir 31, 96, 100, 172
Shuman Canyon 191
Sierra Blanca 117
Sierra Blanca Formation 65, 78, 79, 116, 172
Sierra Madre 13, 16, 77, 87, 93, 106, 127, 194
Sierra Nevada 4, 8, 86, 176
sills 119
Simi Valley 84
Simmler Formation 87, 195, 199, 204, 206
Simonton Cove 44, 46, 121
Sisquoc 131, 186, 188
Sisquoc Formation 18, 51, 65, 66, 90, 123, 124, 125, 130, 150, 165, 168, 178, 179, 182, 187, 189, 191, 193, 194, 198
Sisquoc River 20, 77, 93, 106, 127, 132, 158, 186, 200, 202
Sisquoc River Valley 24, 194
slough
  Barka Slough 130, 165, 190, 191
  Devereux Slough 54
  Goleta Slough 17, 26, 53, 54, 101, 138, 154
slumps 170
Soledad Mountain 119
solfatara 131
Solomon Hills 20, 24, 93, 108, 134

Solvang 22, 86, 127, 142, 158, 169, 178, 186, 187
Sonora, Mexico 116
South Cuyama fault 107, 206, 207
South Cuyama Oil Field 135
South Point Formation 64
Southern California Continental Borderland 4
spheroidal weathering 161, 174
spreading center 1, 60, 143
spring tides 48
springs 32
Stanley Mountain 204
Stearns Wharf 37, 38, 101
strike-slip faults 98
subduction zone 1
submarine canyon 34
submarine seeps 128
Suey Canyon 200
Summerland 88, 107, 134
Summerland Beach 49, 95
Summerland Oil Field 132, 152
Sur-Nacimiento fault 73, 106, 195, 204
Surf 21, 127, 178, 180
Sutil Island 113
Sweeney Road 198
Sweetwater Creek 128
Sycamore Canyon 84, 96, 173, 185
Sycamore Creek 28
syncline 13
 Laurel Canyon syncline 167
 Los Alamos syncline 157, 190, 191
 Painted Cave syncline 167, 175
 San Miguelito syncline 164

T

Tajiguas 89
Tajiguas Creek 155, 158
tar seep 129
Tecololito Creek 54
Tecolote Tunnel 31, 111, 141
Temblor Formation 75, 172, 181
Tepusquet Canyon 16, 127, 194, 199
Tepusquet Quarry 127
terrace deposits 168, 187
thrust fault 98
Tierra del Fuego 2
tombolo 48
Toro Canyon 152
Tranquillon Mountain 9, 13, 18, 58, 74, 90, 117, 178
Tranquillon Formation 58, 65, 178, 185, 188, 197, 201
Transverse Range 4, 5, 14, 67, 86, 97, 98, 103, 145
travertine 142, 188
tunnel 174
 Doulton Tunnel 31
 Gaviota Pass Tunnel 155
 Mission Tunnel 31, 174
 Tecolote Tunnel 31, 111, 141
Twitchell Dam 19, 25
Twitchell Reservoir 31, 202

U

UCSB 54, 108, 112, 131, 154
Uinta Range 4
Upper Oso Forest Service Campground 180, 181
upwelling 92, 125

V

valleys
 Carpinteria Valley 33
 Central Valley 4, 22, 99
 Central Valley of Santa Cruz Island (Cañada del Medio) 21, 22, 114
 Cuddy Valley 104
 Cuyama Valley 8, 24, 87, 97, 106, 194, 206
 Goleta Valley 33, 101
 Honda Valley 169
 Lockwood Valley 105
 Lompoc Valley 178
 Los Alamos Valley 19, 20, 24, 96, 108, 193
 Ojai Valley 142

San Antonio Creek Valley 20
Santa Maria Valley 20, 24, 32, 93, 132, 158, 186
Santa Rita Valley 19
Santa Ynez River Valley 8, 22, 160, 161, 170
Santa Ynez Valley 24, 96, 166, 167, 196
Simi Valley 84
Yosemite Valley 8, 39
Vancouver, George 54, 60, 129
Vandenberg Air Force Base 24, 130, 164, 169, 190, 191, 193
Vandenberg Village 164
Vaqueros Formation 26, 63, 64, 65, 87, 88, 135, 155, 156, 163, 180, 188, 195, 197, 200, 204
Vaqueros sandstone 159
Ventucopa 206
Ventura
Ventura County 9, 24, 25, 34, 44, 84, 85, 86, 99, 100, 102, 104, 105, 106, 108, 114, 119, 120, 127, 131, 134, 136, 142, 151, 173, 206
Ventura River 142
Veronica Springs 29
volcanics
   Santa Cruz Island 63, 115
volcanic rock 61, 120, 131

W

Wagon Flat Forest Service Campground 195
warping 108
water gap 28
wave erosion 8
West Camino Cielo 160, 170, 171
West Cat Canyon Oil Field 134, 192
Western Cordillera 2, 4
White Elephant barite mine 127, 195
White Hills 18, 123
Whiterock Bluff 205
Wildhorse Mountain 127
Winchester Gun Club 161
wind gap 28

Y

Ynezan orogeny 82
Yosemite Valley 8, 39

Z

Zaca Creek 30, 72, 157, 187
Zaca Lake 30, 31, 103, 127, 185
Zaca Peak 169, 185
Zaca Ridge 30
Zanja de Cota Creek 22, 178
zinc 127, 143